Yuval Noah Harari

David Vandermeulen • Daniel Casanave

Sapiens
A Graphic History

The Birth of
Humankind

VOLUME ONE

HARPER PERENNIAL

NEW YORK • LONDON • TORONTO • SYDNEY • NEW DELHI • AUCKLAND

HARPER ● PERENNIAL

Adapted under the coordination of Sapienship and Albin Michel
Publishing (France) from *Sapiens: A Brief History of Humankind*
by Yuval Noah Harari.

First published in Hebrew in 2011 by Kinneret, Zmora-Bitan, Dvir.
First English language edition published in 2020 by HarperCollins
Publishers.

HarperCollins books may be purchased for educational, business,
or sales promotional use. For information, please email the
Special Markets Department at SPsales@harpercollins.com.

First U.S. edition

Creation and co-writing: Yuval Noah Harari
Adaptation and co-writing: David Vandermeulen
Adaptation and illustration: Daniel Casanave
Colors: Claire Champion

Editor (Albin Michel): Martin Zeller

Sapienship Storytelling:
Sponsorship and management: Itzik Yahav
Management and editing: Naama Avital
Editing and coordination: Naama Wartenburg
Master text translation: Adriana Hunter
Diversity consulting: Slava Greenberg

www.sapienship.co

Library of Congress Cataloging-in-Publication Data has been
applied for.

ISBN 978-0-06-305133-1
ISBN 978-0-06-305508-7 (library edition)

20 21 22 23 24   TC   10 9 8 7 6 5 4 3 2

To the extinct, the lost and the forgotten.
Everything that comes together is bound to be dissolved.
—YUVAL NOAH HARARI

TIMELINE OF HISTORY

13.8 BILLION	MATTER AND ENERGY APPEAR. BEGINNING OF PHYSICS. FORMATION OF ATOMS AND MOLECULES. BEGINNING OF CHEMISTRY.
4.5 BILLION	FORMATION OF PLANET EARTH.
3.8 BILLION	EMERGENCE OF ORGANISMS. BEGINNING OF BIOLOGY.
6 MILLION	LAST COMMON GRANDMOTHER OF HUMANS AND CHIMPANZEES.
2.5 MILLION	HUMANS EVOLVE IN AFRICA. FIRST STONE TOOLS.
2 MILLION	HUMANS SPREAD FROM AFRICA TO EURASIA. EVOLUTION OF DIFFERENT HUMAN SPECIES.
500,000	NEANDERTHALS EVOLVE IN EUROPE AND THE MIDDLE EAST.
300,000	DAILY USAGE OF FIRE.
200,000	HOMO SAPIENS EVOLVES IN AFRICA.
70,000	THE COGNITIVE REVOLUTION. EMERGENCE OF STORYTELLING. BEGINNING OF HISTORY. SAPIENS SPREAD OUT OF AFRICA.
50,000	SAPIENS SETTLE AUSTRALIA. EXTINCTION OF AUSTRALIAN MEGAFAUNA.
30,000	EXTINCTION OF NEANDERTHALS. HOMO SAPIENS THE ONLY SURVIVING HUMAN SPECIES.

16,000	SAPIENS SETTLE AMERICA. EXTINCTION OF AMERICAN MEGAFAUNA.
12,000	THE AGRICULTURAL REVOLUTION. DOMESTICATION OF PLANTS AND ANIMALS.
5,000	FIRST KINGDOMS, SCRIPT AND MONEY. POLYTHEISTIC RELIGIONS.
4,250	FIRST EMPIRE—THE AKKADIAN EMPIRE OF SARGON.
2,500	INVENTION OF COINS—A UNIVERSAL MONEY. THE PERSIAN EMPIRE—A UNIVERSAL POLITICAL ORDER. BUDDHISM IN INDIA—A UNIVERSAL TEACHING.
2,000	HAN EMPIRE IN CHINA. ROMAN EMPIRE IN THE MEDITERRANEAN. CHRISTIANITY.
1,400	ISLAM.
500	THE SCIENTIFIC REVOLUTION. HUMANKIND ADMITS ITS IGNORANCE AND BEGINS TO ACQUIRE UNPRECEDENTED POWER. EUROPEANS BEGIN TO CONQUER AMERICA AND THE OCEANS. THE ENTIRE PLANET BECOMES A SINGLE HISTORICAL ARENA. THE RISE OF CAPITALISM.
200	THE INDUSTRIAL REVOLUTION. FAMILY AND COMMUNITY ARE REPLACED BY STATE AND MARKET. MASSIVE EXTINCTION OF PLANTS AND ANIMALS.
THE PRESENT	HUMANS TRANSCEND THE BOUNDARIES OF PLANET EARTH. NUCLEAR WEAPONS THREATEN THE SURVIVAL OF HUMANKIND. ORGANISMS ARE INCREASINGLY SHAPED BY INTELLIGENT DESIGN RATHER THAN NATURAL SELECTION.
THE FUTURE	INTELLIGENT DESIGN BECOMES THE BASIC PRINCIPLE OF LIFE? FIRST NON-ORGANIC LIFE-FORMS? HUMANS BECOME GODS?

REBELS OF
THE SAVANNAH

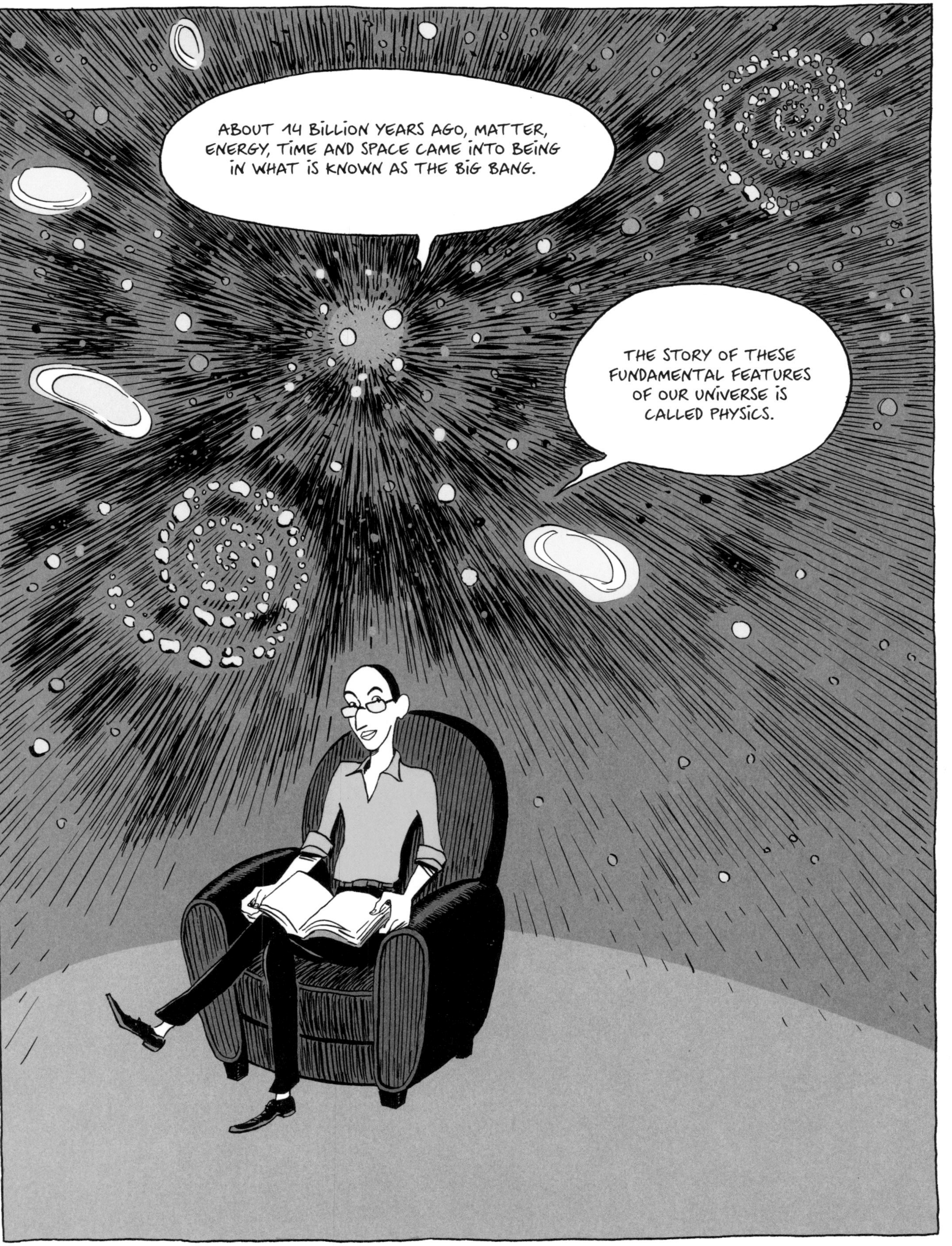
ABOUT 14 BILLION YEARS AGO, MATTER, ENERGY, TIME AND SPACE CAME INTO BEING IN WHAT IS KNOWN AS THE BIG BANG.
THE STORY OF THESE FUNDAMENTAL FEATURES OF OUR UNIVERSE IS CALLED PHYSICS.

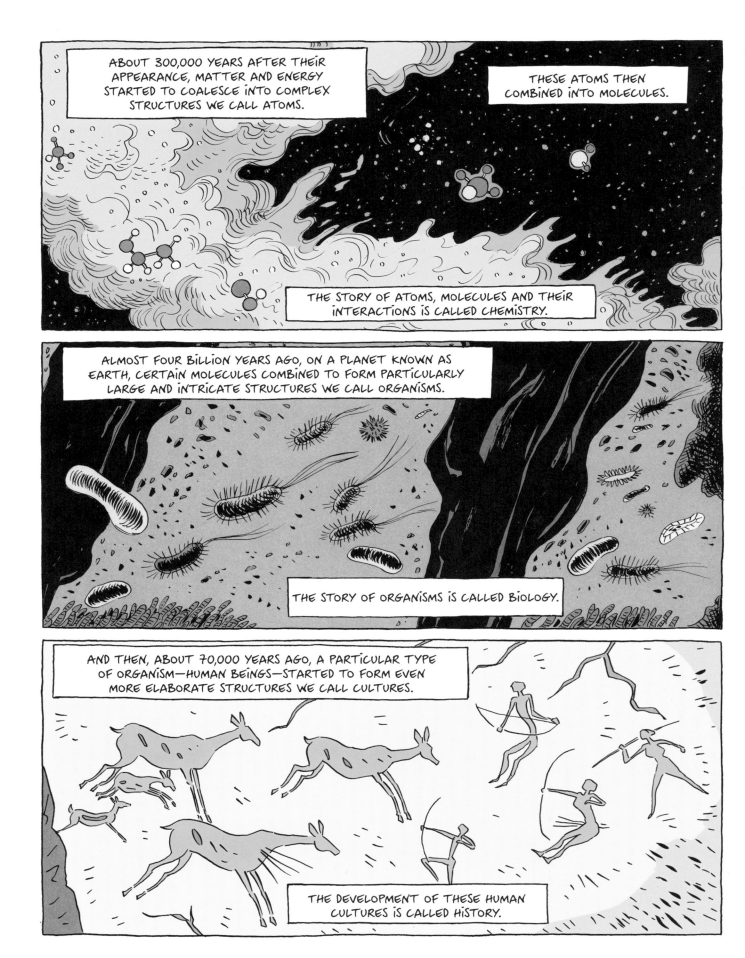

ABOUT 300,000 YEARS AFTER THEIR APPEARANCE, MATTER AND ENERGY STARTED TO COALESCE INTO COMPLEX STRUCTURES WE CALL ATOMS.

THESE ATOMS THEN COMBINED INTO MOLECULES.

THE STORY OF ATOMS, MOLECULES AND THEIR INTERACTIONS IS CALLED CHEMISTRY.

ALMOST FOUR BILLION YEARS AGO, ON A PLANET KNOWN AS EARTH, CERTAIN MOLECULES COMBINED TO FORM PARTICULARLY LARGE AND INTRICATE STRUCTURES WE CALL ORGANISMS.

THE STORY OF ORGANISMS IS CALLED BIOLOGY.

AND THEN, ABOUT 70,000 YEARS AGO, A PARTICULAR TYPE OF ORGANISM—HUMAN BEINGS—STARTED TO FORM EVEN MORE ELABORATE STRUCTURES WE CALL CULTURES.

THE DEVELOPMENT OF THESE HUMAN CULTURES IS CALLED HISTORY.

SORRY, I HAVEN'T EVEN INTRODUCED MYSELF!

HI. MY NAME'S YUVAL NOAH HARARI.

I'M A HISTORIAN.

I KNOW, HISTORIANS DON'T USUALLY TALK ABOUT PHYSICS, CHEMISTRY AND BIOLOGY...

THEY USUALLY TALK ABOUT STUFF LIKE THE FRENCH REVOLUTION.

BUT HUMAN HISTORY IS ACTUALLY A DIRECT CONTINUATION OF PHYSICS...

EINSTEIN

CHEMISTRY...

CURIE

AND BIOLOGY.

DARWIN

WE CAN'T UNDERSTAND THINGS LIKE THE FRENCH REVOLUTION UNTIL WE UNDERSTAND HOW HUMANS EVOLVED.

HUMANS ARE ANIMALS, AND EVERYTHING THAT HAS HAPPENED IN HISTORY HAS OBEYED THE LAWS OF PHYSICS, CHEMISTRY AND BIOLOGY.

ATOMIC ENERGY

CHEMISTRY

MICROSCOPE AND LAB SET

THERE WERE HUMAN BEINGS LONG BEFORE THERE WAS HISTORY. THE FIRST HUMANS EVOLVED RIGHT HERE, IN EAST AFRICA.

2.5 MILLION YEARS AGO

SEE THAT LITTLE GROUP OVER THERE?

THEY'RE OUR ANCIENT ANCESTORS.

THEIR BEHAVIOR LOOKS KIND OF FAMILIAR, DON'T YOU THINK?

BUT THEY WEREN'T VERY DIFFERENT FROM GORILLAS...

ELEPHANTS...

OR BIRDS.

THE TRUTH IS THERE WAS NOTHING SPECIAL ABOUT THESE EARLY HUMANS. THEY WERE STILL REGULAR ANIMALS WITH NO MORE IMPACT ON THEIR ENVIRONMENT THAN BABOONS, FIREFLIES OR JELLYFISH. THERE WAS NO SIGN THAT ONE DAY THEY WOULD CONQUER AND TRANSFORM THE WHOLE WORLD...

12

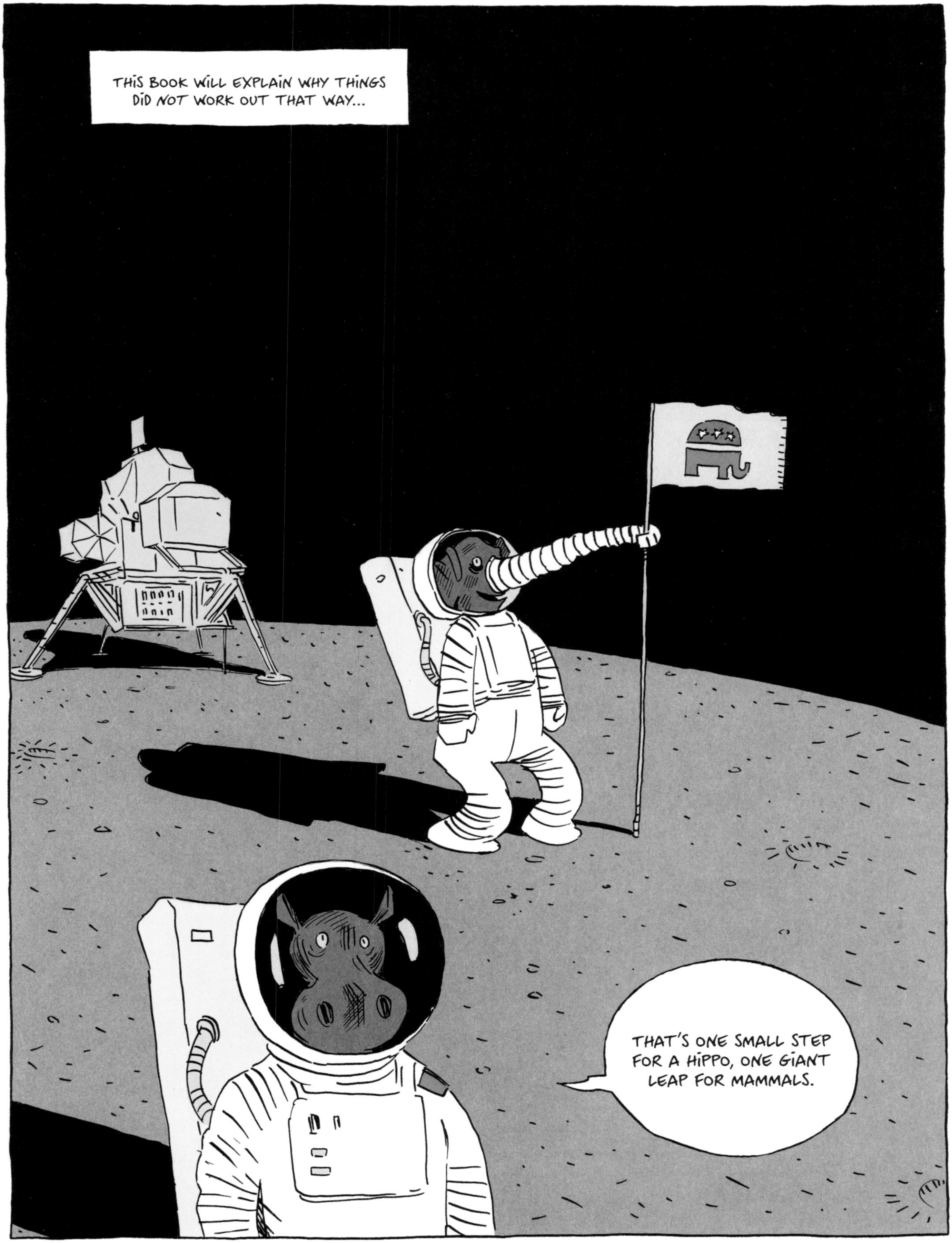
THIS BOOK WILL EXPLAIN WHY THINGS DID NOT WORK OUT THAT WAY...
THAT'S ONE SMALL STEP FOR A HIPPO, ONE GIANT LEAP FOR MAMMALS.

ONE VERY IMPORTANT THING ABOUT ANCIENT HUMANS IS THEY DIDN'T ALL BELONG TO THE SAME SPECIES.

NOWADAYS, PEOPLE AROUND THE WORLD MAY LOOK DIFFERENT AND SPEAK DIFFERENT LANGUAGES, BUT WE'RE ALL THE SAME SPECIES—HOMO SAPIENS.

WHICH IS SURPRISING... AFTER ALL, THERE ARE DIFFERENT SPECIES OF ANTS, SNAKES AND BEARS... SO WHY SHOULDN'T THERE BE DIFFERENT SPECIES OF HUMANS?

IN FACT, UNTIL ABOUT 50,000 YEARS AGO, OUR PLANET WAS HOME TO AT LEAST SIX DIFFERENT SPECIES OF HUMANS. ONE LILLIPUT-LIKE ISLAND WAS INHABITED BY A SPECIES OF DWARVES...

DING DONG!

WHO CAN THAT BE?

EXCUSE ME! SOMEONE'S AT THE DOOR.

HI UNCLE YUVAL!

ZOE! PERFECT TIMING! I'VE GOT SOMETHING TO SHOW YOU.

A CARD GAME?

THAT LOOKS FUN! CAN WE PLAY?

SURE WE CAN!

JUST WAIT, YOU'RE GONNA LOVE IT!

HOMININS

HOMO NEANDERTHALENSIS
NICKNAME: NEANDERTHAL MAN

THE HUMAN FAMILY 50,000 YEARS AGO

- **Location:** Europe and Western Asia   +6
- **Survival:** −300,000 to −30,000   +3
- **Characteristic:** Very big brain, very muscular.   +4
- **Strength:** Resistant to ice-age climate. Looked after the old and the disabled   +6
- **Weakness:** Terrible at public relations. Favorite butt of cartoon jokes.   −1

HOMO ERECTUS
NICKNAME: UPRIGHT MAN

THE HUMAN FAMILY 50,000 YEARS AGO

- **Location:** All over Asia   +5
- **Survival:** −2 million to −50,000   +7
- **Characteristic:** The longest surviving human species   +8
- **Strength:** First to use fire and make hunting weapons...   +4
- **Weakness:** They made the same weapons for hundreds of thousands of years   +1

HOMO LUZONENSIS
NICKNAME: MAN FROM LUZON ISLAND

THE HUMAN FAMILY 50,000 YEARS AGO

- **Location:** Luzon, Philippines
- **Survival:** −700,000 to −50,000   +1
- **Characteristic:** The new kids on the block, discovered only in 2019   +5
- ...   +1
- gth: Adapted to tropical forests   +6
- ...ness: Total lack of public relations.   −2

HOMO DENISOVA
NICKNAME: THE DENISOVANS

THE HUMAN FAMILY 50,000 YEARS AGO

- **Location:** Siberia, East Asia   +3
- **Survival:** −300,000 to −50,000   +2
- **Characteristic:** The mystery man. Originally identified from just one fossilized finger   +2
- **Strength:** Very sociable. Slept around with Sapiens and Neanderthals   +5
- **Weakness:** Stuck their fingers into strange places   −1

HOMO SAPIENS
NICKNAME: WISE MAN

THE HUMAN FAMILY 50,000 YEARS AGO

- **Location:** Everywhere, even on the moon   +10
- **Survival:** −300,000 to right now. May die out by 2100   +4
- **Characteristic:** Think they're smarter than everybody else.   +0
- **Strength:** Good at making tools. From toothbrushes to intercontinental nuclear missiles.   +10
- **Weakness:** Huge ego. Often happy believe nonsense.   −3

HOMO FLORESIENSIS
NICKNAME: FLORES MAN, THE "HOBBI..."

THE HUMAN FAMILY 50,000 YEARS AGO

- **Location:** Island of Flores, Indonesia
- **Survival:** −800,000 to −50,000
- **Characteristic:** Dwarf species. About 1 meter tall and weighing 25 kg.   +1
- **Strength:** Hunted the island's dwarf elephants and giant lizards.   +3
- **Weakness:** Stay at home type. Never left the island.   +2   +4   +1

FASCINATING, ISN'T IT?

BUT, UNCLE YUVAL, ARE YOU TOTALLY SURE ALL THOSE OTHER HUMAN SPECIES DIED OUT AND WE'RE ALL HOMO SAPIENS NOW?

YES, WE KNOW THAT FOR SURE. WE ALL BELONG TO THE SAME SPECIES.

WHEREVER YOU GO TODAY, TO FRANCE, SENEGAL, AUSTRALIA OR GREENLAND, YOU'LL ONLY MEET ONE SPECIES OF HUMAN BEING.

HOW CAN WE BE SO SURE?

HOW CAN YOU TELL IF TWO PEOPLE BELONG TO THE SAME SPECIES OR TWO DIFFERENT SPECIES?

DO YOU KNOW WHAT A SPECIES REALLY IS?

ER... WELL, NOT REALLY, TO BE HONEST!

COME ALONG THEN. LET'S GO FIND AN EXPERT TO ANSWER ALL THESE QUESTIONS!

IS IT FAR?

NOT FAR AT ALL. SHE WORKS RIGHT HERE, AT THE UNIVERSITY

PROFESSOR SARASWATI IS A BIOLOGIST WHO SPECIALIZES IN CLASSIFICATION. SHE'S CRAZY ABOUT CLASSIFYING, ORDERING AND ORGANIZING EVERYTHING.

BIOLOGY DEPARTMENT. PROF. SARASWATI

KNOCK KNOCK!

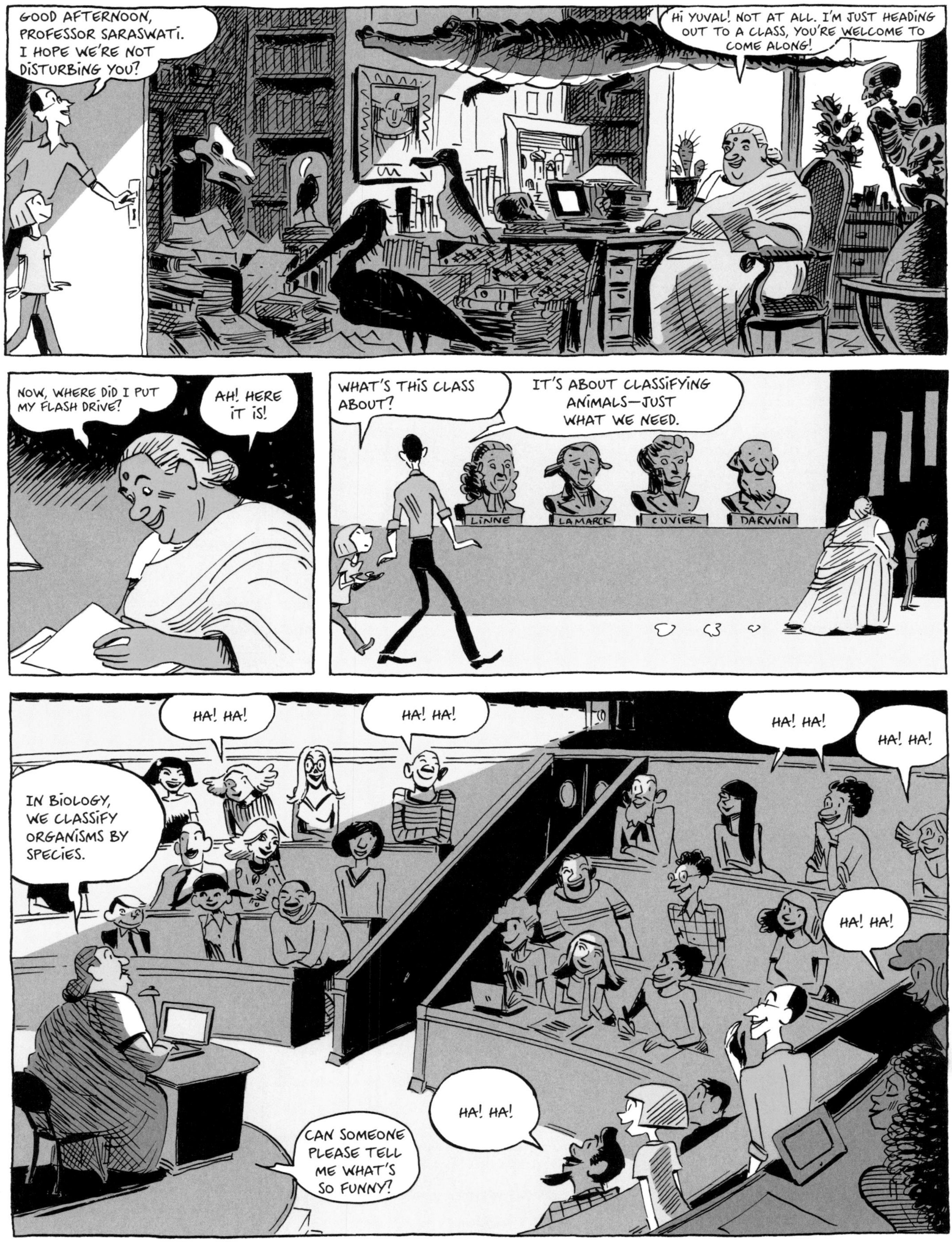

GOOD AFTERNOON, PROFESSOR SARASWATI. I HOPE WE'RE NOT DISTURBING YOU?
Hi YUVAL! NOT AT ALL. I'M JUST HEADING OUT TO A CLASS, YOU'RE WELCOME TO COME ALONG!
NOW, WHERE DID I PUT MY FLASH DRIVE?
AH! HERE IT IS!
WHAT'S THIS CLASS ABOUT?
IT'S ABOUT CLASSIFYING ANIMALS—JUST WHAT WE NEED.
LINNE
LAMARCK
CUVIER
DARWIN
IN BIOLOGY, WE CLASSIFY ORGANISMS BY SPECIES.
HA! HA!
HA! HA!
HA! HA!
HA! HA!
HA! HA!
HA! HA!
CAN SOMEONE PLEASE TELL ME WHAT'S SO FUNNY?

WOOPS! I THINK THAT'S THE WRONG SLIDE!

I THOUGHT I CHECKED THESE... MY SUMMER HOLIDAY SNAPS SHOULDN'T BE IN THIS FILE.

DARN! THEY'RE ALL MIXED UP!

IT DOESN'T MATTER. THIS PHOTO IS ACTUALLY VERY USEFUL.

JUST LET ME WRITE A FEW POINTS ON THE BOARD.

THIS IS ME WITH MY DOG KIKI, ON HER BIRTHDAY.

ME
FAMILY: HOMINIDS
GENUS: HOMO
SPECIES: SAPIENS

KIKI
FAMILY: CANIDS
GENUS: CANIS
SPECIES: CANIS FAMILIARIS (DOG)

CLASSIFYING ANIMALS IS A COMPLEX BUSINESS. AS WELL AS THE SPECIES, THE GENUS AND THE FAMILY, YOU MUST CONSIDER THE SUB-FAMILY, THE INFRA-ORDER, THE SUB-ORDER, THE CLASS, THE PHYLUM.... AND YOU ALSO HAVE TO DEAL WITH HYBRIDIZATION, SUB-SPECIES, MICRO-SPECIES, RING-SPECIES... AND DON'T FORGET

CONFUSING PHENOMENA LIKE CONVERGENT EVOLUTION, MIMICRY, PLESIOMORPHIC FEATURES... IN THE CASE OF HOMO SAPIENS AND CANIS FAMILIARIS, IT IS OBVIOUS THAT THEY ARE DIFFERENT SPECIES, BUT WHAT ABOUT CANIS FAMILIARIS AND CANIS LUPUS? SINCE DOGS AND WOLVES CAN INTERBREED, SOME SCIENTISTS CLASSIFY THEM AS THE SAME SPECIES, CALLING THE DOG CANIS LUPUS FAMILIARIS. OTHER SCIENTISTS DISAGREE, ARGUING THAT THE REAL CRITERION FOR DEFINING THE BORDER BETWEEN RELATED SPECIES IS NOT INTERBREEDING, BUT RATHER THE STATISTICS OF GENE FLOWS...

I DIDN'T REALLY GET ALL THAT. SCIENCE IS WAY COMPLICATED...

IT SURE IS! I'M LOST TOO...

EXCUSE ME, PROFESSOR, COULD YOU SIMPLIFY THINGS A LITTLE SO THAT EVEN A HISTORIAN CAN FOLLOW WHAT YOU'RE SAYING? IT WOULD HELP OTHER PEOPLE TOO, LIKE MY NIECE...

OH, FORGIVE ME! I FORGOT ABOUT YOU TWO. OF COURSE, I'LL TRY TO MAKE IT SIMPLER!

NOW WE'LL GET SOME ANSWERS!

SO, AS I WAS SAYING... APART FROM THE FACT THAT WE'RE BOTH MAMMALS, ANYONE CAN SEE THAT KIKI AND I DON'T HAVE A WHOLE LOT IN COMMON. WE'RE NOT THE SAME SPECIES. WE DON'T EVEN BELONG TO THE SAME FAMILY!

ON THE OTHER HAND, THIS DONKEY AND THIS HORSE HAVE A LOT OF CHARACTERISTICS IN COMMON. IN FACT, THEY SHARE A RECENT ANCESTOR, THE HIPPIDION.

HIPPIDION

DONKEY

HORSE

ANIMALS BELONG TO THE SAME SPECIES IF THEY TEND TO MATE NATURALLY AND PRODUCE FERTILE OFFSPRING. BUT THIS MARE AND THIS DONKEY DON'T SEEM TO BE THAT INTO EACH OTHER...

THEY WILL MATE IF YOU INDUCE THEM, BUT THEY'LL PRODUCE MULES AND HINNIES, WHICH ARE STERILE.

THE MULE: THE RESULT OF MATING A JACK DONKEY WITH A MARE.

THE HINNY: THE RESULT OF MATING A STALLION WITH A JENNY DONKEY.

THIS MEANS THAT MUTATIONS IN DONKEY DNA CAN'T CROSS OVER TO HORSES, OR VICE VERSA. SO THESE TWO ANIMALS ARE CONSIDERED TWO DISTINCT SPECIES, MOVING ALONG SEPARATE EVOLUTIONARY PATHS. AS MORE MUTATIONS OCCUR, THEY'LL BECOME INCREASINGLY DIFFERENT.

AH! IT'S KIKI AGAIN— ANOTHER SLIP-UP... BUT IT'S JUST WHAT WE NEED!

SEE HOW DIFFERENT KIKI AND THIS BULLDOG LOOK. YET THEY'RE MEMBERS OF THE SAME SPECIES AND THEY SHARE THE SAME DNA POOL.

THEY WILL HAPPILY MATE, AND THEIR PUPPIES WILL GROW UP TO PAIR OFF WITH OTHER DOGS AND HAVE PUPPIES OF THEIR OWN.

CLOSELY RELATED SPECIES THAT EVOLVED FROM A COMMON ANCESTOR ARE BUNCHED TOGETHER UNDER THE HEADING "GENUS"

LION

TIGER

PANTHERA

LEOPARD

JAGUAR

WE BIOLOGISTS LABEL ORGANISMS WITH A TWO-PART LATIN NAME, FIRST THE GENUS THEN THE SPECIES. LIONS, FOR EXAMPLE, ARE CALLED *PANTHERA LEO*, THE SPECIES LEO OF THE GENUS PANTHERA.

PANTHERA LEO

ALL THE PEOPLE IN THIS AMPHITHEATER ARE HOMO SAPIENS: WE'RE OF THE GENUS HOMO, WHICH MEANS "HUMAN," AND THE SPECIES SAPIENS WHICH MEANS "WISE."

IN TODAY'S WORLD, PEOPLE MAY LOOK VERY DIFFERENT. SOME HAVE DARK SKIN, OTHERS HAVE LIGHT SKIN. SOME HAVE STRAIGHT HAIR, OTHERS HAVE CURLY HAIR. SOME ARE VERY TALL, OTHERS ARE SHORT.

BUT PEOPLE FROM ALL OVER THE PLANET CAN HAVE SEX TOGETHER AND PRODUCE FERTILE CHILDREN.

SO, THEY'RE ALL THE SAME "SPECIES."

ALL THE DISTINCTIONS THAT SEEM SO IMPORTANT TODAY— FRENCH AND GERMAN, CHRISTIAN AND MUSLIM, BLACK AND WHITE—ARE VERY RECENT INVENTIONS, AND THEY DON'T HAVE MUCH INFLUENCE ON HUMAN EVOLUTION.

50,000 YEARS AGO THERE WERE NO FRENCH OR GERMAN, CHRISTIAN OR MUSLIM, BLACK OR WHITE.

TODAY, A WHITE CHRISTIAN GERMAN MAN CAN BE IN A SEXUAL RELATIONSHIP WITH A BLACK MUSLIM FRENCH WOMAN, AND HAVE CHILDREN WHO WILL PASS ON THEIR COMBINED GENETIC CODES TO FUTURE GENERATIONS.

WE'RE NOW THE ONLY REMAINING SPECIES OF THE GENUS HOMO. BUT THERE WERE ONCE MANY OTHERS, LIKE THE DWARF HOMO FLORESIENSIS. THEIR NAME MEANS "HUMAN FROM FLORES ISLAND."

SEX BETWEEN HOMO SAPIENS AND HOMO FLORESIENSIS MAY HAVE BEEN POSSIBLE, BUT IT WAS UNLIKELY TO RESULT IN FERTILE OFFSPRING...

SO, WE ORGANIZE ANIMALS INTO SPECIES AND GENERA...

...THEN WE GROUP THE GENERA INTO FAMILIES.

THE CAT FAMILY

THE DOG FAMILY

THE ELEPHANT FAMILY

ALL MEMBERS OF A PARTICULAR FAMILY CAN TRACE THEIR LINEAGE BACK TO A FOUNDING MATRIARCH OR PATRIARCH.

PROAILURUS

FOR EXAMPLE, THE SMALL HOUSECAT AND THE GIANT LION HAVE A COMMON FELINE ANCESTOR—THE PROAILURUS—THAT LIVED ABOUT 25 MILLION YEARS AGO.

HUMANS BELONG TO A FAMILY TOO—THE FAMILY OF GREAT APES.

MAN   CHIMPANZEE   GORILLA   ORANGUTAN

HUMANS BELONG TO THE FAMILY OF GREAT APES...

A FEW CENTURIES AGO, THE PROFESSOR WOULD HAVE BEEN BURNED AT THE STAKE FOR SAYING THAT. AND THIS BOOK WOULD DEFINITELY HAVE BEEN GIVEN THE SAME TREATMENT...

THE FACT THAT HUMANS HAVE A FAMILY USED TO BE ONE OF HISTORY'S MOST CLOSELY GUARDED SECRETS. FOR A LONG TIME HOMO SAPIENS LIKED TO SEE THEMSELVES AS SET APART FROM OTHER ANIMALS. AN ORPHAN WITH NO FAMILY, NO BROTHERS OR SISTERS, NO COUSINS

AND DEFINITELY NO PARENTS.

BUT LIKE IT OR NOT, WE HAVE A BIG—AND NOISY!—FAMILY.

OUR CLOSEST RELATIVES ARE THE CHIMPANZEES. THIS COULD BE WHAT HAPPENED, SIX MILLION YEARS AGO.

WHAT BEAUTIFUL CHILDREN YOU HAVE...

SIX MILLION YEARS AGO, THIS FEMALE APE HAD TWO DAUGHTERS.

ONE IS THE ANCESTOR OF ALL CHIMPANZEES.

AND THIS ONE IS OUR GRANDMOTHER...

PREVIOUSLY, IN
SEASON 1
MULTICELLULAR ORGANISMS INVENT SEX FOR THE FIRST TIME!

SEASON 6

PREVIOUSLY, IN
SEASON 3
EXTINCTION OF THE DINOSAURS!

THAT FIRST KISS WAS UNFORGETTABLE!

EVOLUTION!
THE GREATEST SHOW ON EARTH!

EVERYBODY THOUGHT T. REX WOULD SLAY THE COMPETITION THIS SEASON! BUT NOBODY SAW THAT ASTEROID COMING, RIGHT?

GUESS WHAT? I GOT TO BE THE WINNER!

APPROXIMATELY TWO MILLION YEARS AGO.

WELCOME TO THE NEW SEASON OF EVOLUTION, THE GREATEST SHOW ON EARTH!

YOU ARE OUR NINE COUPLES OF HUMAN ADVENTURERS WHO'VE RISEN TO THE CHALLENGE OF LEAVING YOUR HOME IN AFRICA TO COLONIZE NEW, UNCHARTED TERRITORIES. TO SUCCEED, YOU'LL HAVE TO ADAPT TO DIFFERENT CLIMATES, LANDSCAPES AND FOOD!

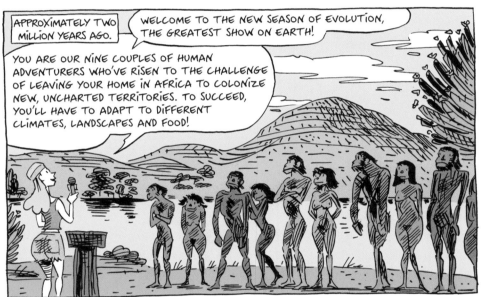

THE FIRST EPISODE BEGINS IN YOUR BACKYARD, EAST AFRICA.

YOU ALL START ON AN EQUAL FOOTING—YOU'RE ALL HUMAN BEINGS.

YOU'VE ALL MADE THE BRAVE DECISION TO LEAVE FAMILIAR TERRITORY BEHIND. BUT, MAKE NO MISTAKE, BY THE END OF THIS SEASON YOU WILL HAVE CHANGED. THE MANY CHALLENGES YOU ENCOUNTER WILL ALTER YOUR DNA, YOUR BODY AND YOUR BRAIN. GENERATION AFTER GENERATION, YOUR LINEAGE WILL EVOLVE IN RESPONSE TO CHANGES IN FOOD AND CLIMATE... YOU WILL BECOME DIFFERENT SPECIES.

EVERYONE READY? ON YOUR MARK, GET SET, GO!

JOIN US AGAIN IN TWO MILLION YEARS, TO SEE WHO WILL WIN THIS NEW SEASON OF EVOLUTION, THE GREATEST SHOW ON EARTH!!!

OUCH! WE ALREADY HAVE OUR FIRST ACCIDENT!

SO TELL ME, LUCY, WHAT HAPPENED?

OH DEAR, THAT'S A NASTY SPRAIN, YOUR POOR ANKLE!

WELL, I'M AFRAID THE ADVENTURE'S OVER FOR YOU ALREADY, LUCY. GUESS YOU'LL BE STAYING HERE IN ETHIOPIA...

IS THAT REALLY HOW IT HAPPENED, UNCLE YUVAL?

NO, NOT EXACTLY. DON'T BELIEVE EVERYTHING YOU SEE ON TV... THIS SHOW MAKES EVOLUTION LOOK LIKE A COMPETITION BETWEEN COUPLES. IN REAL LIFE THE FOUNDING POPULATION OF A SPECIES WAS ALWAYS MUCH LARGER. HUMANS LIVED IN GROUPS RATHER THAN PAIRS. I GUESS THE SHOW DOESN'T HAVE THE BUDGET FOR THAT....

AND THE FAMOUS LUCY DOESN'T BELONG HERE AT ALL! SHE WAS AN AUSTRALOPITHECUS. SHE DIED A MILLION YEARS BEFORE THE FIRST HUMANS APPEARED...

AND NOW, A SHORT BREAK FOR ADS. DON'T GO AWAY! WE'LL BE RIGHT BACK WITH YOU IN A MILLION YEARS!

WHO WILL WIN

SEASON 6

OF

EVOLUTION!

THE GREATEST SHOW ON EARTH!

THE HUNKY NEANDERTHALS?

THE CUNNING DWARVES FROM FLORES ISLAND?

THE MYSTERIOUS DENISOVANS FROM SIBERIA?

STAY WITH US TO FIND OUT!

I ALREADY KNOW HOW THE SEASON ENDS! HOMO SAPIENS WINS AND ALL THE OTHER SPECIES DISAPPEAR!

SPOILER ALERT

THIS FAMOUS PICTURE IS SO MISLEADING! IT MAKES IT LOOK LIKE THERE WAS NEVER MORE THAN ONE HUMAN SPECIES ON EARTH AT A TIME!

THIS WOULD BE NEARER THE TRUTH.

SO HOW COME THERE'S ONLY ONE HUMAN SPECIES TODAY?

IT'S VERY STRANGE... MAYBE EVEN SUSPICIOUS. WE SAPIENS ARE TOO COMFORTABLE WITH BEING THE ONLY HUMAN SPECIES AROUND...

MAYBE WE HAVE GOOD REASONS TO REPRESS MEMORIES OF OUR BROTHERS AND SISTERS.

CEMETERY OF THE HUMAN DYNASTY

HOMO
ERECTUS
2 MILLION
– 50,000

HOMO
DENISOVENSIS
– 300,000
– 50,000

HOMO
FLORESIENSIS
– 800,000
– 50,000

OMG, THESE WERE ALL MEMBERS OF OUR FAMILY?!

OH YES, THEY WERE ALL HUMANS...

HOMO
NEANDERTHALENSIS
– 300,000
– 30,000

WHEN WE SAY "HUMAN" TODAY, WE ONLY MEAN HOMO SAPIENS. BECAUSE WE DON'T REMEMBER ALL THE OTHER HUMAN SPECIES. IT WOULD BE MORE ACCURATE TO CALL OURSELVES "SAPIENS."

THE DIFFERENT HUMAN SPECIES WERE PRETTY DISTINCT FROM ONE ANOTHER. SOME WERE VERY BIG, OTHERS WERE DWARVES. BUT THEY HAD SEVERAL CHARACTERISTICS IN COMMON. ESPECIALLY THEIR BRAINS— ALL HUMANS HAVE EXTRAORDINARILY LARGE BRAINS.

OH YEAH?!

26

YOU CAN'T SEE JUST BY LOOKING AT THEM, BUT IF YOU COMPARE THE BRAINS OF ANIMALS OF ROUGHLY THE SAME WEIGHT, IT'S REALLY STRIKING!

400 CM³ 64 CM³ 122 CM³ 900 CM³ 1500 CM³

CHIMPANZEE | GREAT DANE | SHEEP | HOMO ERECTUS | NEANDERTHAL

THE BRAIN OF MODERN SAPIENS IS VERY IMPRESSIVE TOO: 1300 CM³!

OBVIOUSLY! BECAUSE WE GET SMARTER AND SMARTER ALL THE TIME, RIGHT?

THAT'S ONLY PARTLY TRUE. THE ANCIENT NEANDERTHALS HAD A BIGGER BRAIN THAN OURS. IT WAS ON AVERAGE 1500 CM³!

AND IF LARGE BRAINS ARE A NORMAL DEVELOPMENT, HOW COME OTHER ANIMALS DIDN'T EVOLVE THEM?

YES, HOW COME? I THINK HAVING A BIG BRAIN'S A GOOD IDEA...

SURE, BUT IF YOU HAVE A BIG BRAIN, YOU NEED A BIG SKULL! AND A BIG HEAD'S NOT THAT EASY TO CART AROUND!

ALSO, A BIG BRAIN NEEDS LOTS OF ENERGY... YOU HAVE TO FEED IT EVERY DAY!

AND WHAT'S THE PAYBACK? WHAT'S THE POINT OF BIG BRAINS? WE DON'T REALLY KNOW...

IT MAKES SENSE NOW THAT WE LIVE IN MODERN CITIES. BUT ON THE SAVANNA?

A STORY IN MY COMIC BOOK SAYS THE EXACT SAME THING!

PREHISTORIC BILL

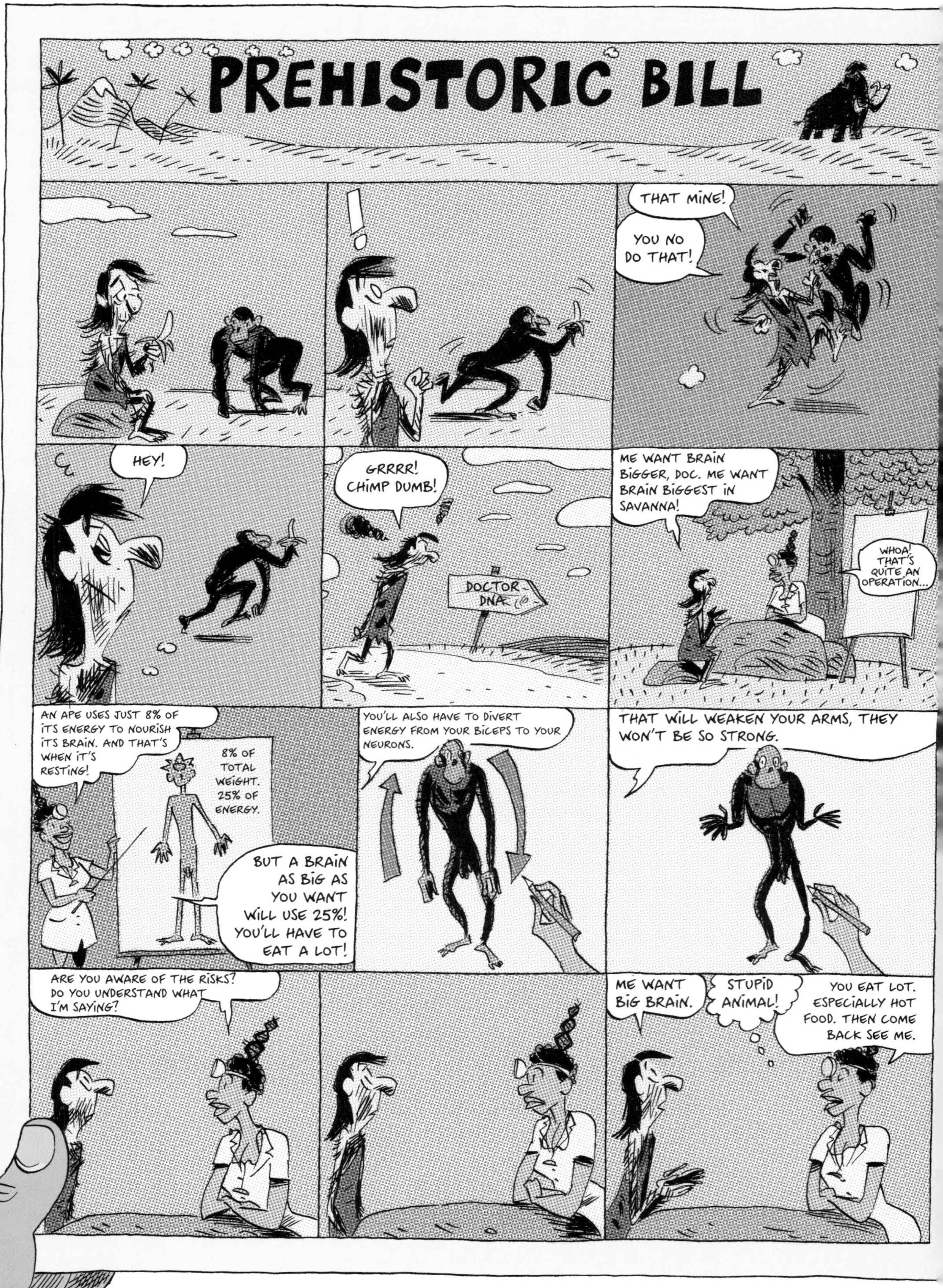
PREHISTORIC BILL
!
THAT MINE!
YOU NO DO THAT!
HEY!
GRRRR! CHIMP DUMB!
DOCTOR DNA
ME WANT BRAIN BIGGER, DOC. ME WANT BRAIN BIGGEST IN SAVANNA!
WHOA! THAT'S QUITE AN OPERATION...
AN APE USES JUST 8% OF ITS ENERGY TO NOURISH ITS BRAIN. AND THAT'S WHEN IT'S RESTING!
8% OF TOTAL WEIGHT. 25% OF ENERGY.
BUT A BRAIN AS BIG AS YOU WANT WILL USE 25%! YOU'LL HAVE TO EAT A LOT!
YOU'LL ALSO HAVE TO DIVERT ENERGY FROM YOUR BICEPS TO YOUR NEURONS.
THAT WILL WEAKEN YOUR ARMS, THEY WON'T BE SO STRONG.
ARE YOU AWARE OF THE RISKS? DO YOU UNDERSTAND WHAT I'M SAYING?
ME WANT BIG BRAIN.
STUPID ANIMAL!
YOU EAT LOT. ESPECIALLY HOT FOOD. THEN COME BACK SEE ME.

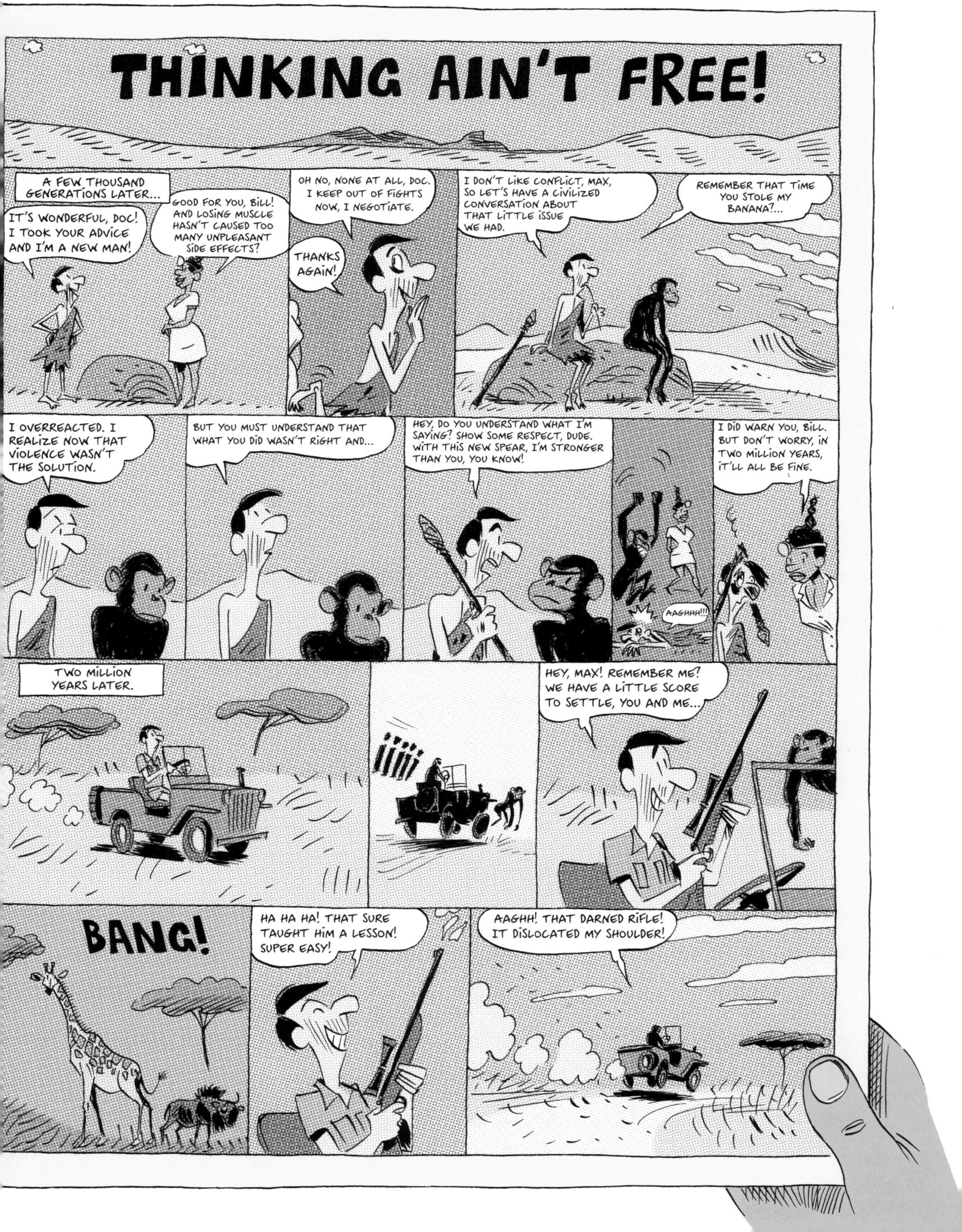
THINKING AIN'T FREE!
A FEW THOUSAND GENERATIONS LATER...
IT'S WONDERFUL, DOC! I TOOK YOUR ADVICE AND I'M A NEW MAN!
GOOD FOR YOU, BILL! AND LOSING MUSCLE HASN'T CAUSED TOO MANY UNPLEASANT SIDE EFFECTS?
OH NO, NONE AT ALL, DOC. I KEEP OUT OF FIGHTS NOW, I NEGOTIATE.
THANKS AGAIN!
I DON'T LIKE CONFLICT, MAX, SO LET'S HAVE A CIVILIZED CONVERSATION ABOUT THAT LITTLE ISSUE WE HAD.
REMEMBER THAT TIME YOU STOLE MY BANANA?...
I OVERREACTED. I REALIZE NOW THAT VIOLENCE WASN'T THE SOLUTION.
BUT YOU MUST UNDERSTAND THAT WHAT YOU DID WASN'T RIGHT AND...
HEY, DO YOU UNDERSTAND WHAT I'M SAYING? SHOW SOME RESPECT, DUDE. WITH THIS NEW SPEAR, I'M STRONGER THAN YOU, YOU KNOW!
AAGHHH!!!
I DID WARN YOU, BILL. BUT DON'T WORRY, IN TWO MILLION YEARS, IT'LL ALL BE FINE.
TWO MILLION YEARS LATER.
HEY, MAX! REMEMBER ME? WE HAVE A LITTLE SCORE TO SETTLE, YOU AND ME...
BANG!
HA HA HA! THAT SURE TAUGHT HIM A LESSON! SUPER EASY!
AAGHH! THAT DARNED RIFLE! IT DISLOCATED MY SHOULDER!

HMM, VERY FUNNY!

YES, BUT I DON'T LIKE THE ENDING!

NOWADAYS WE CAN TRAVEL FAST THANKS TO OUR CARS...

...AND FIGHT WITH OUR RIFLES, BUT THESE ARE ALL QUITE RECENT ACHIEVEMENTS.

FOR TWO MILLION YEARS, HUMAN NEURAL NETWORKS JUST KEPT ON GROWING, BUT APART FROM SOME FLINT KNIVES AND POINTED STICKS, ANCIENT HUMANS HAD VERY LITTLE TO SHOW FOR ALL THAT BRAIN POWER.

SO, WHAT WAS DRIVING THE EVOLUTION OF OUR HUGE BRAINS FOR THOSE TWO MILLION YEARS?

FRANKLY, WE DON'T KNOW.

YET ANOTHER INTERESTING OPEN QUESTION. I'LL BORROW THAT, YUVAL, FOR MY LIST OF FUTURE RESEARCH PROJECTS.

WHAT DROVE THE EVOLUTION OF OUR HUGE BRAINS FOR TWO MILLION YEARS?

ANOTHER UNIQUE HUMAN TRAIT IS THAT WE WALK UPRIGHT, ON TWO LEGS!

IT HAD ITS ADVANTAGES. STANDING UP MADE IT EASIER TO SCAN THE SAVANNA FOR GAME OR ENEMIES.

AND WE NO LONGER NEEDED OUR ARMS FOR WALKING SO WE COULD USE THEM FOR OTHER THINGS

LIKE THROWING STONES, SIGNALING OR CRADLING BABIES.

THE MORE THINGS ANCIENT HUMANS DID WITH THEIR HANDS, THE MORE THEIR HANDS EVOLVED! AN INCREDIBLY SOPHISTICATED NETWORK OF FINELY TUNED NERVES AND MUSCLES DEVELOPED IN THE PALMS AND FINGERS.

THAT'S HOW HUMANS CAN NOW PERFORM VERY INTRICATE TASKS WITH THEIR HANDS.

SUCH AS MAKING AND USING COMPLEX TOOLS. TOOLS ARE THE MAIN CRITERIA ARCHAEOLOGISTS USE TO IDENTIFY ANCIENT HUMANS. THE EARLIEST TOOLS DATE BACK ABOUT 2.5 MILLION YEARS!

BUT THERE'S ALWAYS A PRICE TO PAY! WALKING UPRIGHT HAD ITS DOWNSIDE.

OUR PRIMATE ANCESTORS HAD SMALL HEADS AND THEY WALKED ON ALL FOURS!

ADAPTING TO AN UPRIGHT POSITION WAS A CHALLENGE, ESPECIALLY WHEN THE BACK AND NECK HAD TO SUPPORT AN EXTRA-LARGE HEAD.

THAT'S WHY PEOPLE SO OFTEN HAVE BACKACHE OR A STIFF NECK.

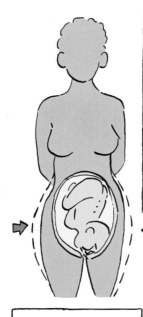

WOMEN PAID EXTRA. WALKING UPRIGHT REQUIRED NARROWER HIPS, AND THAT CONSTRICTED THE BIRTH CANAL—JUST WHEN BABIES' HEADS KEPT GETTING BIGGER.

DEATH IN CHILDBIRTH BECAME A MAJOR RISK FOR MOTHERS. WOMEN WHO GAVE BIRTH EARLIER, WHEN THE BABY'S BRAIN AND HEAD WERE STILL RELATIVELY SMALL AND SUPPLE, HAD A BETTER CHANCE OF SURVIVING.

SO NATURAL SELECTION FAVORED EARLIER BIRTHS.

I CAN TELL YOU THAT CHILDBIRTH IS THE MOST PAINFUL THING I EVER EXPERIENCED. I HAVE THREE CHILDREN, AND WITH EACH BIRTH I ASKED MYSELF: IS IT WORTH IT, JUST TO WALK UPRIGHT AND HAVE A BIGGER BRAIN?!

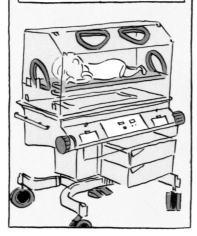

COMPARED TO OTHER ANIMALS, HUMANS ARE BORN PREMATURELY.

A CALF CAN TROT SHORTLY AFTER BIRTH, BUT HUMAN BABIES ARE HELPLESS AND THEY'RE DEPENDENT ON THEIR ELDERS FOR FOOD, PROTECTION AND EDUCATION FOR MANY YEARS.

RAISING CHILDREN REQUIRED CONSTANT HELP FROM OTHER FAMILY MEMBERS AND NEIGHBORS. IT TAKES A TRIBE TO RAISE A HUMAN. SO, EVOLUTION FAVORED THE ONES WHO COULD FORM STRONG SOCIAL TIES.

PARADOXICALLY, THE HELPLESSNESS OF HUMAN BABIES TURNED OUT TO BE A BLESSING. IT MEANT HUMANS HAD TO DEVELOP THEIR SOCIAL SKILLS.

BECAUSE HUMANS ARE BORN UNDERDEVELOPED, THEY CAN BE EDUCATED AND SOCIALIZED FAR MORE THAN ANY OTHER ANIMAL.

MOST MAMMALS EMERGE FROM THE WOMB LIKE GLAZED EARTHENWARE EMERGING FROM A KILN.

ANY ATTEMPT AT REMOLDING THEM WILL ONLY SCRATCH OR BREAK THEM.

HUMANS COME OUT OF THE WOMB LIKE MOLTEN GLASS. THEY CAN BE SPUN, STRETCHED AND SHAPED WITH A SURPRISING DEGREE OF FREEDOM.

WE ASSUME THAT LARGE BRAINS, USING TOOLS AND HAVING GREAT LEARNING ABILITIES AND COMPLEX SOCIAL STRUCTURES ARE HUGE ADVANTAGES. BUT HUMANS ENJOYED ALL THESE ADVANTAGES FOR TWO MILLION YEARS AND WERE STILL WEAK, MARGINAL CREATURES.

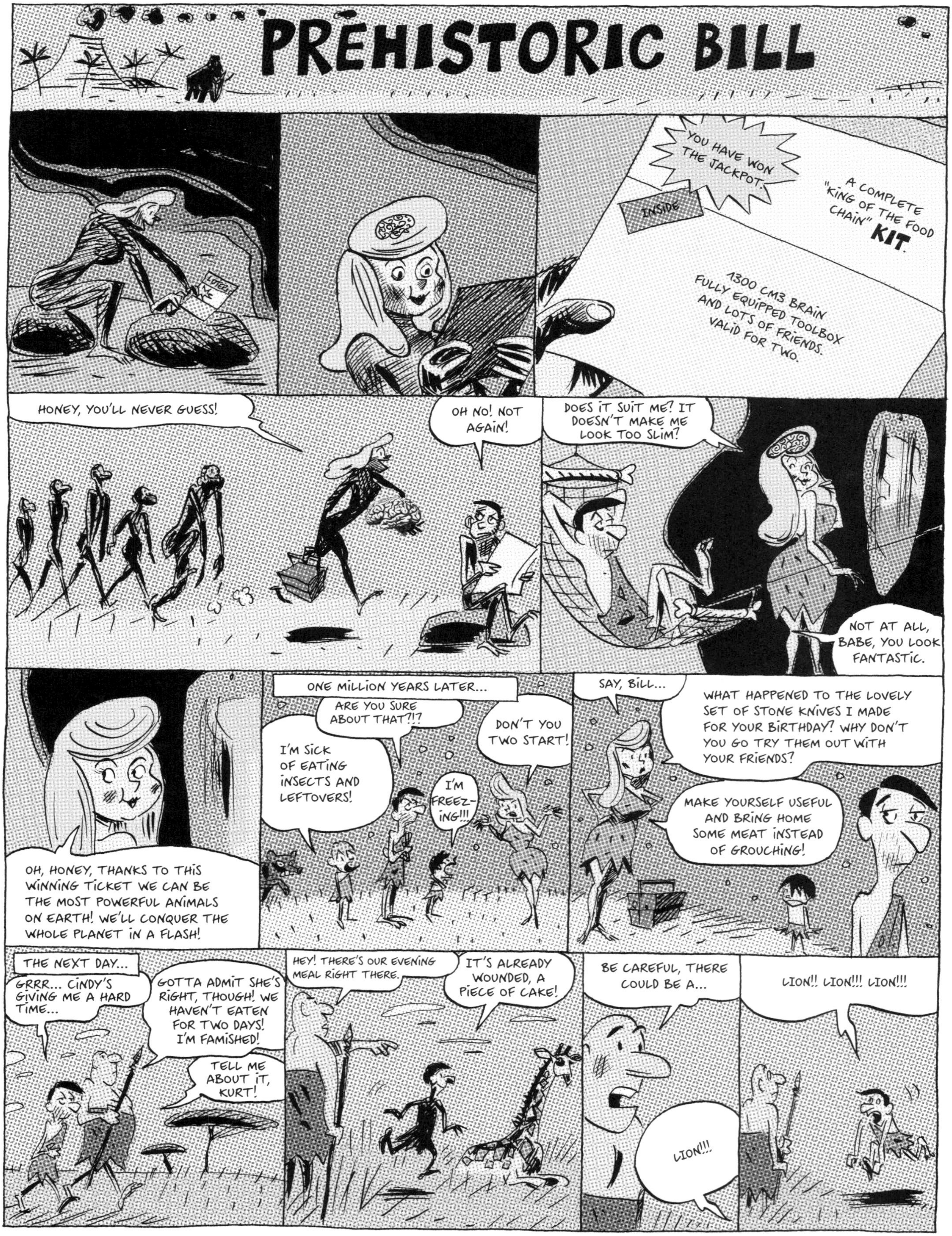
PREHISTORIC BILL
YOU HAVE WON THE JACKPOT.
INSIDE
A COMPLETE "KING OF THE FOOD CHAIN" KIT.
1300 CM3 BRAIN
FULLY EQUIPPED TOOLBOX
AND LOTS OF FRIENDS.
VALID FOR TWO.
HONEY, YOU'LL NEVER GUESS!
OH NO! NOT AGAIN!
DOES IT SUIT ME? IT DOESN'T MAKE ME LOOK TOO SLIM?
NOT AT ALL, BABE, YOU LOOK FANTASTIC.
OH, HONEY, THANKS TO THIS WINNING TICKET WE CAN BE THE MOST POWERFUL ANIMALS ON EARTH! WE'LL CONQUER THE WHOLE PLANET IN A FLASH!
ONE MILLION YEARS LATER...
I'M SICK OF EATING INSECTS AND LEFTOVERS!
ARE YOU SURE ABOUT THAT?!?
I'M FREEZ-ING!!!
DON'T YOU TWO START!
SAY, BILL...
WHAT HAPPENED TO THE LOVELY SET OF STONE KNIVES I MADE FOR YOUR BIRTHDAY? WHY DON'T YOU GO TRY THEM OUT WITH YOUR FRIENDS?
MAKE YOURSELF USEFUL AND BRING HOME SOME MEAT INSTEAD OF GROUCHING!
THE NEXT DAY...
GRRR... CINDY'S GIVING ME A HARD TIME...
GOTTA ADMIT SHE'S RIGHT, THOUGH! WE HAVEN'T EATEN FOR TWO DAYS! I'M FAMISHED!
TELL ME ABOUT IT, KURT!
HEY! THERE'S OUR EVENING MEAL RIGHT THERE.
IT'S ALREADY WOUNDED, A PIECE OF CAKE!
BE CAREFUL, THERE COULD BE A...
LION!!!
LION!! LION!!! LION!!!

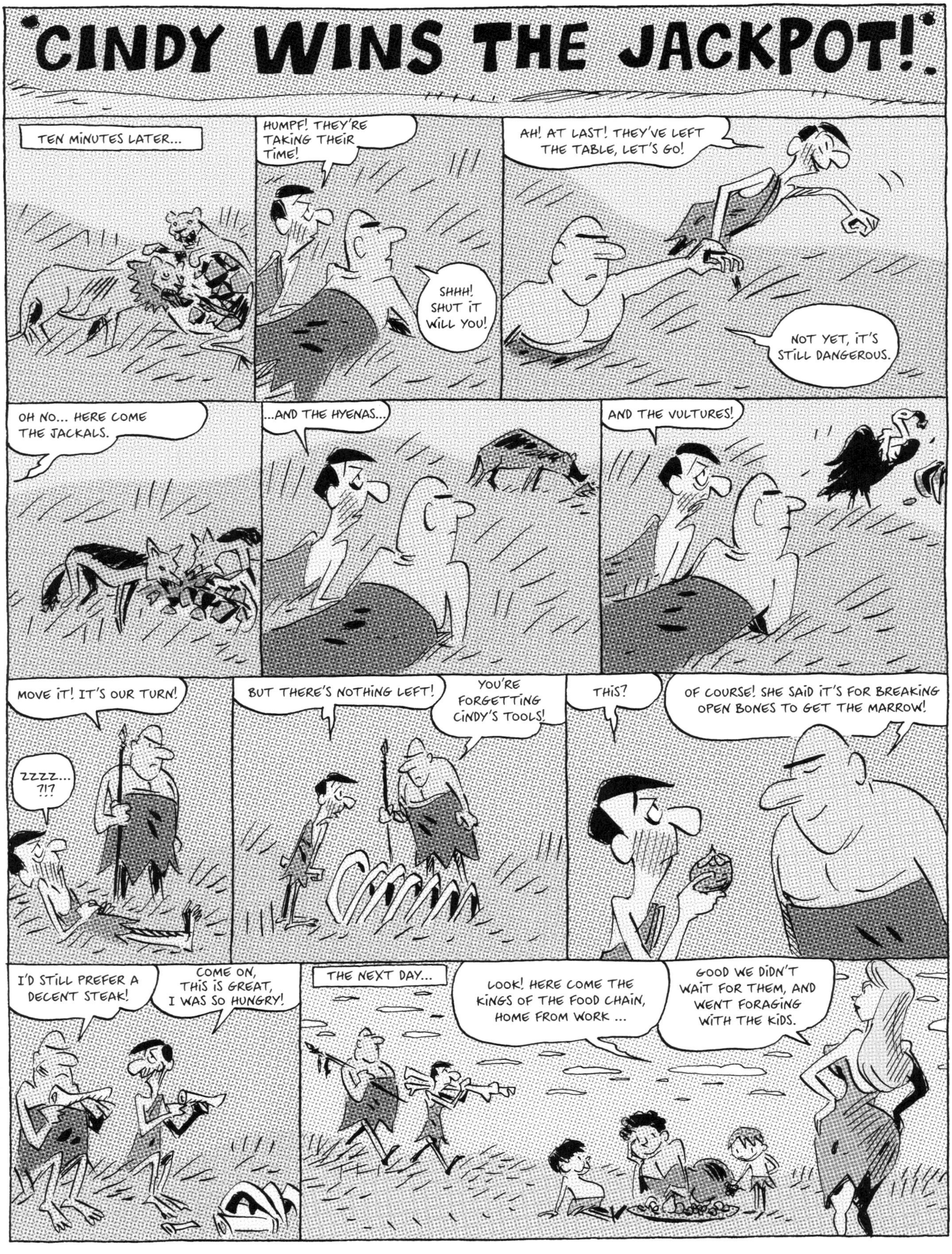
CINDY WINS THE JACKPOT!
TEN MINUTES LATER...
HUMPF! THEY'RE TAKING THEIR TIME!
SHHH! SHUT IT WILL YOU!
AH! AT LAST! THEY'VE LEFT THE TABLE, LET'S GO!
NOT YET, IT'S STILL DANGEROUS.
OH NO... HERE COME THE JACKALS.
...AND THE HYENAS...
AND THE VULTURES!
MOVE IT! IT'S OUR TURN!
ZZZZ... ?!?!!!
BUT THERE'S NOTHING LEFT!
YOU'RE FORGETTING CINDY'S TOOLS!
THIS?
OF COURSE! SHE SAID IT'S FOR BREAKING OPEN BONES TO GET THE MARROW!
I'D STILL PREFER A DECENT STEAK!
COME ON, THIS IS GREAT, I WAS SO HUNGRY!
THE NEXT DAY...
LOOK! HERE COME THE KINGS OF THE FOOD CHAIN, HOME FROM WORK ...
GOOD WE DIDN'T WAIT FOR THEM, AND WENT FORAGING WITH THE KIDS.

THAT'S RIGHT! EVEN WITH THEIR TOOLS, THEIR AMAZING ABILITY TO LEARN AND THEIR COMPLEX SOCIAL STRUCTURES, HUMANS WERE JUST WEAK, MARGINAL CREATURES FOR A GOOD TWO MILLION YEARS!

ZOO

←TICKETS

NO BAGS

SEVERAL HUMAN SPECIES BEGAN HUNTING LARGE GAME ABOUT 400,000 YEARS AGO.

BUT HOMO SAPIENS DIDN'T REACH THE TOP OF THE FOOD CHAIN TILL THE LAST 100,000 YEARS.

UNTIL RECENTLY, WE WERE INSIGNIFICANT ANIMALS SOMEWHERE IN THE MIDDLE OF THE FOOD CHAIN. THEN WE SUDDENLY JUMPED TO THE TOP.

MAYBE TOO SUDDENLY. IT TOOK LIONS, EAGLES AND SHARKS MILLIONS OF YEARS TO GRADUALLY REACH THE TOP OF THE PYRAMID.

ANCESTOR OF THE GAZELLE

THIS GRADUAL PROCESS MEANT THE ECOSYSTEM COULD DEVELOP CHECKS AND BALANCES TO PREVENT LIONS AND SHARKS FROM WREAKING TOO MUCH HAVOC.

AS THE LIONS GREW DEADLIER, GAZELLES EVOLVED TO RUN FASTER.

ANCESTOR OF THE HYENA

HYENAS LEARNED TO COOPERATE WITH ONE ANOTHER.

ANCESTOR OF THE RHINOCEROS

RHINOCEROSES BECAME MORE FEROCIOUS.

BUT HUMANKIND JUMPED TO THE TOP SO QUICKLY THAT THE ECOSYSTEM DIDN'T HAVE TIME TO ADJUST.

EVEN HUMANS THEMSELVES DIDN'T ADJUST.

REALLY! HOW COME! I THOUGHT WE WERE AT THE TOP BECAUSE NATURE MADE SUCH A GREAT JOB OF US...

HO! HO! TAKE A LOOK AT THE WORLD'S GREAT PREDATORS. MOST OF THEM ARE MAGNIFICENT CREATURES.

MILLIONS OF YEARS OF DOMINION HAVE FILLED THEM WITH SELF-CONFIDENCE.

SAPIENS IS MORE LIKE SOME UPSTART DICTATOR WHO'S ALWAYS AFRAID OF LOSING POWER.

NOT SO LONG AGO, WE WERE THE UNDERDOGS OF THE SAVANNA...

OUR FIRST TOOLS WERE USED TO SCAVENGE WHAT WAS LEFT BY MAJESTIC LIONS AND FEROCIOUS HYENAS.

AND THAT REALLY HELPS US UNDERSTAND OUR HISTORY AND OUR PSYCHOLOGY.

IT EXPLAINS WHY WE'RE SO STRESSED AND ALWAYS PANICKING ABOUT OUR POSITION.

WHICH MAKES US DOUBLY CRUEL AND DANGEROUS. A LOT OF HISTORICAL CALAMITIES, FROM DEADLY WARS TO ECOLOGICAL DISASTERS, RESULTED FROM OUR SUDDEN JUMP TO THE TOP.

WHETHER YOU'RE AN ERECTUS, A NEANDERTHAL OR JUST A SAPIENS, THIS IS FOR YOU!

LIGHT THE FLAME!

A MODERN TECHNIQUE FOR THE WHOLE FAMILY, A BREEZE FOR MEN AND WOMEN ALIKE, AND FOR KIDS OVER TEN! NO MORE CHANCE FIRES. CHOOSE

THE DAILY FLAME! ™

800,000 YEARS AGO, SOME HUMAN SPECIES ONLY HAD FIRES WHEN THE OPPORTUNITY AROSE, BUT THIS IS—300,000, TIME TO GET MODERN!

I CHOSE THE DAILY FLAME™ AND MY WHOLE LIFE HAS CHANGED!

DON'T WAIT FOR RANDOM FIRES PRODUCED BY LIGHTNING. LEARN TO MAKE YOUR OWN FIRE!

PICK YOUR METHOD

A) FRICTION

B) PERCUSSION

AMAZING BENEFITS TO THRILL THE WHOLE FAMILY!

1) NIGHTTIME LIGHTING

WHERE ON EARTH ARE MY SLIPPERS?

THEY'RE UNDER THE SOFA!

2) A WARM COZY HOME!

I'M FREEZING!

AAAH...

3) AN EFFECTIVE DETERRENT TO UNWELCOME GUESTS!

4) MOOD LIGHTING FOR SPECIAL OCCASIONS

HAPPY BIRTHDAY!

I THINK THE EXHIBITION CONTINUES THIS WAY!

5) AND LAST BUT NOT LEAST... AN ASSET FOR YOUR PANTRY! FIRE CAN TRAP GAME (BOTH LARGE AND SMALL), AND TURNS THOSE IMPASSABLE THICKETS INTO A FANTASTIC GIANT BARBECUE!

*SPECIAL STONE AGE BONUS! ONCE THE BUSHFIRE'S EXTINGUISHED, DISCOVER THE DELIGHTS OF HOT FOOD. ROASTED NUTS, POTATOES AND MEAT... MMM... TRULY DELICIOUS!

YES, I WANT TO OPT IN FOR **THE DAILY FLAME** ™ AND WOULD LIKE TO RECEIVE A BROCHURE.

BROCHURE 1

"LEARNING TO MAKE FIRE WITH FRICTION"

PLUS MY INTRODUCTORY GIFT

BROCHURE 2

"LEARNING TO MAKE FIRE WITH PERCUSSION"

PLUS MY INTRODUCTORY GIFT.

NAME.....................

ADDRESS ....

....

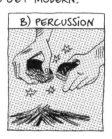

*OFFER AVAILABLE ONLY TO NEANDERTHALS AND SAPIENS.

THE AD DOESN'T MENTION ANOTHER ADVANTAGE— FIRE OPENED THE FIRST SIGNIFICANT GULF BETWEEN HUMANS AND OTHER ANIMALS.

ALMOST ALL ANIMALS DEPEND ON THEIR OWN BODIES FOR THEIR POWER. ON THE STRENGTH OF THEIR MUSCLES, THE SIZE OF THEIR TEETH, THE BREADTH OF THEIR WINGS.

SOME ANIMALS CAN HARNESS WINDS AND CURRENTS BUT THEY CAN'T CONTROL THESE NATURAL FORCES, AND THEY'RE ALWAYS CONSTRAINED BY THEIR OWN PHYSIQUE.

TAKE EAGLES: THEY IDENTIFY THERMAL COLUMNS RISING FROM THE GROUND, SPREAD THEIR GIANT WINGS AND LET THE HOT AIR LIFT THEM UPWARD.

BUT EAGLES CAN'T CONTROL WHERE THESE THERMALS ARE, AND THEIR WINGSPAN DICTATES HOW HIGH OR FAR THEY'LL BE CARRIED.

WHEN HUMANS DOMESTICATED FIRE, THEY GAINED CONTROL OF A MANAGEABLE AND POTENTIALLY LIMITLESS FORCE. UNLIKE EAGLES, HUMANS COULD CHOOSE WHEN AND WHERE TO USE THIS FORCE, AND COULD EXPLOIT FIRE FOR ANY NUMBER OF TASKS.

MORE IMPORTANTLY, THE POWER OF FIRE WASN'T LIMITED BY THE HUMAN BODY'S CONSTRAINTS. A SINGLE HUMAN BEING WITH A FLINT COULD BURN DOWN AN ENTIRE FOREST IN A MATTER OF HOURS.

THE DOMESTICATION OF FIRE WAS A SIGN OF THINGS TO COME. IT WAS THE FIRST IMPORTANT STEP ON THE WAY TO THE ATOM BOMB...

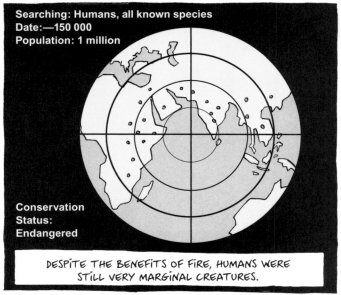

Searching: Humans, all known species
Date:—150 000
Population: 1 million

Conservation
Status:
Endangered

DESPITE THE BENEFITS OF FIRE, HUMANS WERE STILL VERY MARGINAL CREATURES.

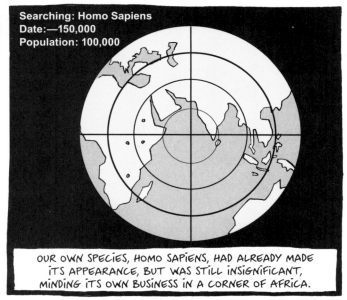

Searching: Homo Sapiens
Date:—150,000
Population: 100,000

OUR OWN SPECIES, HOMO SAPIENS, HAD ALREADY MADE ITS APPEARANCE, BUT WAS STILL INSIGNIFICANT, MINDING ITS OWN BUSINESS IN A CORNER OF AFRICA.

WE DON'T KNOW EXACTLY WHEN ANIMALS THAT CAN BE CLASSIFIED AS HOMO SAPIENS FIRST EVOLVED FROM AN EARLIER TYPE OF HUMANS.

BUT MOST SCIENTISTS AGREE THAT BY 150,000 YEARS AGO, SAPIENS IN EAST AFRICA LOOKED A LOT LIKE US.

I CAN CONFIRM THAT.

VERY MODERN LOOKING. SMALL TEETH AND JAWS BUT A HUGE BRAIN, THE SAME SIZE AS OURS.

DEFINITELY DUE TO THE ADVANTAGES OF FIRE.

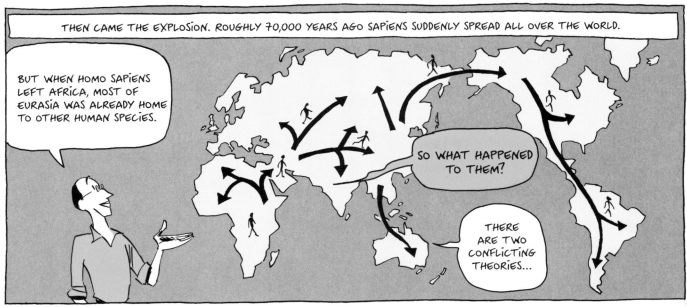

THEN CAME THE EXPLOSION. ROUGHLY 70,000 YEARS AGO SAPIENS SUDDENLY SPREAD ALL OVER THE WORLD.

BUT WHEN HOMO SAPIENS LEFT AFRICA, MOST OF EURASIA WAS ALREADY HOME TO OTHER HUMAN SPECIES.

SO WHAT HAPPENED TO THEM?

THERE ARE TWO CONFLICTING THEORIES...

THE INTERBREEDING THEORY
A WONDERFUL STORY OF ATTRACTION, SEX AND MINGLING.

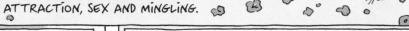

HOMO SAPIENS ARE AFRICAN IMMIGRANTS WHO SPREAD ALL OVER THE WORLD FRATERNIZING WITH OTHER HUMAN POPULATIONS. PRESENT DAY HUMANS ARE THE RESULT OF THIS MELTING POT.

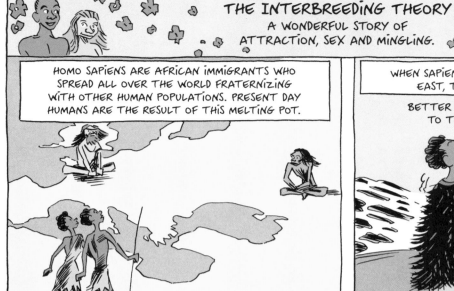

WHEN SAPIENS REACHED EUROPE AND THE MIDDLE EAST, THEY MET THE NEANDERTHALS.

BETTER RESISTANCE TO THE COLD

BIGGER BRAINS

MORE MUSCULAR

NEANDERTHALS ARE OFTEN CARICATURED AS ARCHETYPICAL BRUTISH, STUPID "CAVE PEOPLE," BUT RECENT EVIDENCE HAS CHANGED THEIR IMAGE.

YOU BUILT YOUR BRAND ON BRUTE FORCE FOR 200 YEARS. WAKE-UP CALL! THIS IS THE TWENTY-FIRST CENTURY! HOW ABOUT SOFTENING THE IMAGE A BIT?

YES, THAT WORKS FOR US. WE NEVER REALLY LIKED DRAGGING FEMALES AROUND BY THE HAIR. WE'RE BOTH GAY, ANYWAY.

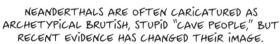

ARCHAEOLOGISTS HAVE DISCOVERED THE BONES OF NEANDERTHALS WHO LIVED FOR MANY YEARS WITH PHYSICAL HANDICAPS. CLEARLY, NEANDERTHALS CARED FOR THEIR SICK AND DISABLED.

USES TOOLS AND FIRE

GOOD AT HUNTING

ACCORDING TO THE INTERBREEDING THEORY, WHEN SAPIENS SPREAD INTO NEANDERTHAL LANDS, THEY INTERBRED UNTIL THE TWO POPULATIONS MERGED. IF THIS IS THE CASE, TODAY'S EURASIANS ARE NOT PURE SAPIENS BUT A BLEND OF SAPIENS AND NEANDERTHALS.

HOMO SAPIENS

HOMO NEANDERTHALENSIS

SIMILARLY, WHEN SAPIENS REACHED EAST ASIA, THEY INTERBRED WITH THE LOCAL ERECTUS, SO SOME MODERN PEOPLE ARE A MIXTURE OF SAPIENS AND ERECTUS.

HOMO SAPIENS

HOMO ERECTUS

DID YOU SEE DIANA'S NEW BOYFRIEND? A REGULAR NEANDERTHAL, LIKE TOTALLY!

THE REPLACEMENT THEORY
A TERRIBLE STORY OF INCOMPATIBILITY, REVULSION AND GENOCIDE.

THIS THEORY STRESSES THAT SAPIENS WAS ANATOMICALLY DIFFERENT FROM OTHER HUMANS. AND MOST LIKELY HAD DIFFERENT MATING HABITS. EVEN DIFFERENT BODY ODORS...

THEY WOULD HAVE HAD LITTLE SEXUAL INTEREST IN EACH OTHER.

EVEN IF A NEANDERTHAL ROMEO AND A SAPIENS JULIET FELL IN LOVE, THEY COULDN'T PRODUCE FERTILE CHILDREN, BECAUSE THE GENETIC GULF SEPARATING THEIR TWO POPULATIONS WAS ALREADY UNBRIDGEABLE.

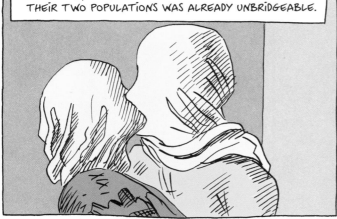

THE TWO POPULATIONS REMAINED COMPLETELY DISTINCT, AND WHEN THE NEANDERTHALS DIED OUT, OR WERE KILLED OFF, THEIR GENES DIED WITH THEM.

ACCORDING TO THIS THEORY, SAPIENS REPLACED ALL PREVIOUS HUMAN POPULATIONS WITH NO INTERBREEDING.

IF THAT'S THE CASE, ALL CONTEMPORARY HUMANS CAN BE TRACED BACK, EXCLUSIVELY, TO EAST AFRICA, 70,000 YEARS AGO. WE'RE ALL "PURE SAPIENS."

WHAT'S YOUR VIEW ON THESE TWO THEORIES, PROFESSOR SARASWATI?

THERE'S A LOT AT STAKE HERE, JANICE.

ARYA SARASWATI

HOMO SAPIENS

SEX-ADDICT OR SERIAL KILLER?

IF THE REPLACEMENT THEORY IS CORRECT, ALL 7.6 BILLION HUMAN BEINGS LIVING ON EARTH TODAY HAVE ROUGHLY THE SAME GENETIC BAGGAGE. WE ALL CAME OUT OF AFRICA IN THE LAST 70,000 YEARS—AND FROM AN EVOLUTIONARY PERSPECTIVE, THAT'S NOT A LONG TIME.

AND IT WOULD CONFIRM THAT RACIAL DISTINCTIONS ARE NEGLIGIBLE.

EXACTLY!

BUT IF THE INTERBREEDING THEORY IS RIGHT, THERE MIGHT WELL BE GENETIC DIFFERENCES BETWEEN AFRICANS, EUROPEANS AND ASIANS THAT GO BACK HUNDREDS OF THOUSANDS OF YEARS...

THIS CONCEPT IS POLITICAL DYNAMITE! IT COULD PROVIDE MATERIAL FOR EXPLOSIVE RACIAL THEORIES!

SURELY THE REPLACEMENT THEORY'S BEEN WIDELY ACCEPTED FOR A LONG TIME?

YES, BECAUSE IT HAD FIRM ARCHAEOLOGICAL BACKING...

...AND IT WAS POLITICALLY EXPEDIENT TO REJECT THE POSSIBILITY OF SIGNIFICANT GENETIC DIVERSITY BETWEEN MODERN HUMAN POPULATIONS.

WE RESEARCHERS HAD NO DESIRE TO OPEN THE PANDORA'S BOX OF RACISM. I'M SURE YOU UNDERSTAND.

OF COURSE...

HOWEVER, THAT BOX WAS OPENED RECENTLY. IN 2010 SCIENTISTS MANAGED TO COLLECT AND MAP DNA FROM NEANDERTHAL BONES, AND COMPARE IT TO THE DNA OF CONTEMPORARY HUMANS.

THE RESULTS STUNNED THE SCIENTIFIC COMMUNITY!

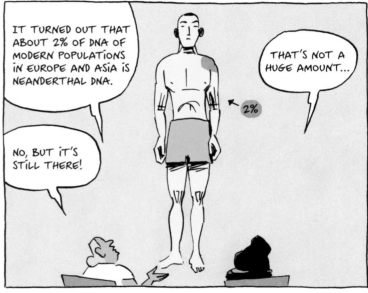

IT TURNED OUT THAT ABOUT 2% OF DNA OF MODERN POPULATIONS IN EUROPE AND ASIA IS NEANDERTHAL DNA.

THAT'S NOT A HUGE AMOUNT...

NO, BUT IT'S STILL THERE!

2%

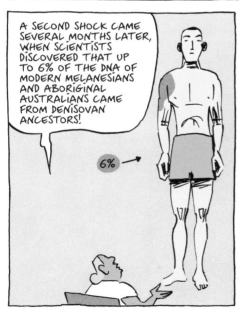

A SECOND SHOCK CAME SEVERAL MONTHS LATER, WHEN SCIENTISTS DISCOVERED THAT UP TO 6% OF THE DNA OF MODERN MELANESIANS AND ABORIGINAL AUSTRALIANS CAME FROM DENISOVAN ANCESTORS!

6%

IF THESE RESULTS TURN OUT TO BE VALID, THE INTERBREEDERS GOT AT LEAST SOME THINGS RIGHT.

SO THE REPLACEMENT THEORY WOULD BE TOTALLY WRONG.

WELL... NOT EXACTLY.

I'M JUST SAYING...

WHICH THEORY IS CORRECT?

☐ A. THE INTERBREEDING THEORY.

☐ B. THE REPLACEMENT THEORY.

☒ C. IT'S COMPLICATED.

SINCE NEANDERTHALS AND DENISOVANS CONTRIBUTED ONLY A SMALL AMOUNT OF DNA TO OUR PRESENT-DAY GENOME, IT IS IMPOSSIBLE TO SPEAK OF A "MERGER," BETWEEN SAPIENS AND OTHER HUMAN SPECIES.

THE DIFFERENCES BETWEEN THEM WEREN'T LARGE ENOUGH TO COMPLETELY PREVENT FERTILE INTERCOURSE, BUT SUCH CONTACTS WERE VERY RARE.

RIGHT, BUT WHAT EXACTLY WAS THE BIOLOGICAL RELATIONSHIP BETWEEN SAPIENS, NEANDERTHALS AND DENISOVANS?

CLEARLY, THEY WEREN'T COMPLETELY DIFFERENT SPECIES LIKE HORSES AND DONKEYS.

SO THEY WERE DIFFERENT POPULATIONS OF THE SAME SPECIES, LIKE YOUR BULLDOG, BRUTUS, AND YOUR LITTLE POMERANIAN, KIKI?

NO... MORE THAN THAT.

BIOLOGY ISN'T BLACK AND WHITE. THERE ARE IMPORTANT GRAY AREAS.

TWO SPECIES THAT EVOLVED FROM A COMMON ANCESTOR, SUCH AS HORSES AND DONKEYS, WERE AT ONE TIME JUST TWO POPULATIONS OF THE SAME SPECIES, LIKE POMERANIANS AND BULLDOGS.

THERE MUST HAVE BEEN A POINT WHEN THE TWO POPULATIONS WERE ALREADY QUITE DISTINCT, BUT STILL CAPABLE ON RARE OCCASIONS OF HAVING SEX AND PRODUCING FERTILE OFFSPRING.

THEN ANOTHER MUTATION SEVERED THIS LAST CONNECTING THREAD, AND THEY WENT THEIR SEPARATE EVOLUTIONARY WAYS.

IT SEEMS THAT ABOUT 50,000 YEARS AGO, SAPIENS, NEANDERTHALS AND DENISOVANS WERE AT THIS BORDERLINE POINT. THEY WERE ALMOST, BUT NOT QUITE, SEPARATE SPECIES.

IF WE HAVE A FEW MINUTES, I'D LIKE TO EXPLAIN HOW SAPIENS WERE ALREADY VERY DIFFERENT FROM NEANDERTHALS AND DENISOVANS, NOT JUST GENETICALLY AND PHYSICALLY, BUT MORE SIGNIFICANTLY IN THEIR COGNITIVE AND SOCIAL ABILITIES...

THIS IS ALL FASCINATING, PROFESSOR, BUT OUR AIRTIME IS UP.

WHAT A PITY...

WELL, I'LL WRAP THIS UP BY REITERATING THAT IT WAS STILL POSSIBLE, ON RARE OCCASIONS, FOR A SAPIENS AND A NEANDERTHAL TO PRODUCE FERTILE OFFSPRING.

SO THE POPULATIONS DIDN'T MERGE, BUT A FEW LUCKY NEANDERTHAL GENES DID HITCH A RIDE ON THE SAPIENS EXPRESS.

THANK YOU, PROFESSOR SARASWATI!

IT'S DISTURBING—AND PERHAPS THRILLING—TO THINK THAT WE SAPIENS COULD AT ONE TIME HAVE SEX WITH AN ANIMAL FROM A DIFFERENT SPECIES, AND PRODUCE CHILDREN TOGETHER.

If you'd like to marry a Neanderthal
dial 1-800-Neanderthal

SHE'S ONE HELL OF A TEACHER!

and tell us what you think!

BUT IF THE NEANDERTHALS, DENISOVANS AND OTHER HUMAN SPECIES DIDN'T MERGE WITH SAPIENS, HOW COME THEY DISAPPEARED?

A VALLEY IN THE BALKANS, WHERE NEANDERTHALS LIVED FOR HUNDREDS OF THOUSANDS OF YEARS.

ONE POSSIBLE EXPLANATION IS THAT HOMO SAPIENS DROVE THEM TO EXTINCTION.

WITH THEIR SUPERIOR TECHNOLOGY AND SOCIAL SKILLS, SAPIENS WERE MORE EFFICIENT HUNTERS AND GATHERERS.

SO, THEY MULTIPLIED AND SPREAD.

THE LESS RESOURCEFUL NEANDERTHALS FOUND IT INCREASINGLY DIFFICULT TO FEED THEMSELVES. THEIR POPULATION DWINDLED AND THEY SLOWLY DIED OUT.

EXCEPT PERHAPS FOR ONE OR TWO WHO JOINED THEIR SAPIENS NEIGHBORS.

LOOK, GUYS! I FOUND THESE TWO KIDS!

ANOTHER POSSIBILITY IS THAT COMPETITION FOR RESOURCES FLARED UP INTO VIOLENCE AND GENOCIDE.

TOLERANCE ISN'T A SAPIENS TRADEMARK...

IN MODERN TIMES, JUST A SMALL DIFFERENCE IN SKIN COLOR, DIALECT OR RELIGION CAN PROMPT ONE GROUP OF SAPIENS TO EXTERMINATE ANOTHER.

WHY SHOULD ANCIENT SAPIENS HAVE BEEN ANY MORE TOLERANT?

IT MAY WELL BE THAT WHEN SAPIENS ENCOUNTERED NEANDERTHALS, HISTORY SAW ITS FIRST AND MOST SIGNIFICANT ETHNIC-CLEANSING CAMPAIGN.

WHATEVER HAPPENED, THE NEANDERTHALS—ALONG WITH THE OTHER HUMAN SPECIES—POSE ONE OF HISTORY'S GREATEST "WHAT IF" QUESTIONS.

WHO KNOWS HOW THINGS MIGHT HAVE TURNED OUT IF NEANDERTHALS OR DENISOVANS HAD SURVIVED ALONGSIDE HOMO SAPIENS?

YOU'RE NOT HUMAN...

NO PARADISE FOR YOU!

WE HOLD THESE TRUTHS TO BE SELF-EVIDENT, THAT ALL HUMAN SPECIES ARE CREATED EQUAL.

WORKERS OF THE GENUS HOMO, UNITE!

OVER THE LAST 30,000 YEARS, HOMO SAPIENS HAS GROWN SO ACCUSTOMED TO BEING THE ONLY HUMAN SPECIES THAT IT'S HARD FOR US TO CONCEIVE OF ANY OTHER POSSIBILITY. HAVING NO SIBLINGS MAKES IT EASIER FOR US TO IMAGINE WE'RE THE PINNACLE OF CREATION, SEPARATED FROM THE REST OF THE ANIMAL KINGDOM BY AN UNBRIDGEABLE CHASM.

PLANET OF THE APES

WHEN CHARLES DARWIN EXPLAINED THAT HOMO SAPIENS WAS JUST ANOTHER KIND OF ANIMAL, PEOPLE WERE OUTRAGED.

EVEN TODAY, MANY REFUSE TO BELIEVE IT.

WE'LL NEVER FIT IN THERE!

YOU'RE RIGHT, BOB, LET'S GO HOME...

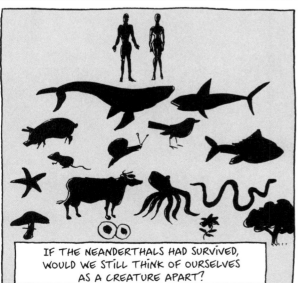

IF THE NEANDERTHALS HAD SURVIVED, WOULD WE STILL THINK OF OURSELVES AS A CREATURE APART?

PERHAPS THIS IS EXACTLY WHY OUR ANCESTORS WIPED OUT THE NEANDERTHALS. THEY WERE TOO FAMILIAR TO IGNORE, BUT TOO DIFFERENT TO TOLERATE.

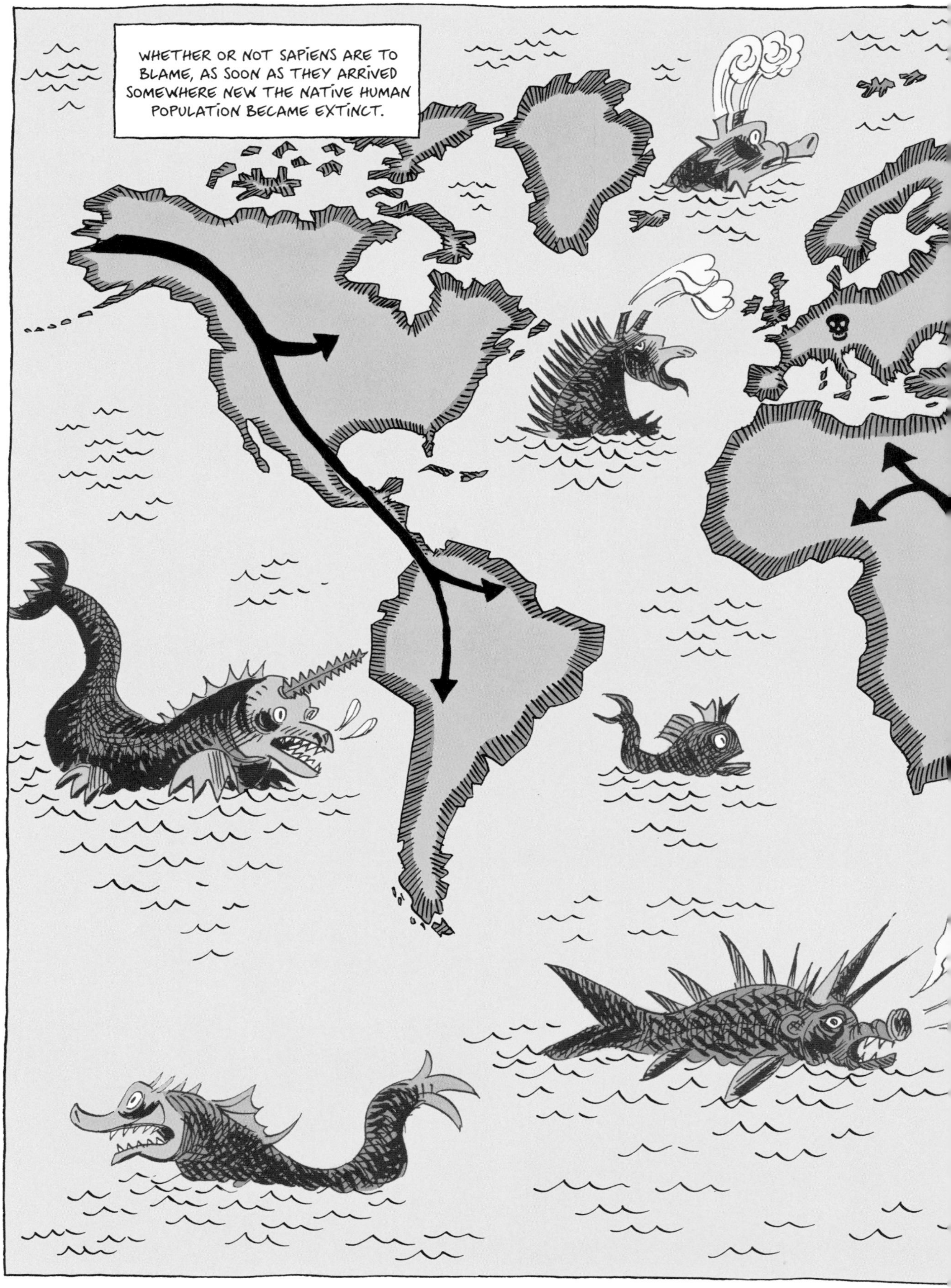
WHETHER OR NOT SAPIENS ARE TO BLAME, AS SOON AS THEY ARRIVED SOMEWHERE NEW THE NATIVE HUMAN POPULATION BECAME EXTINCT.

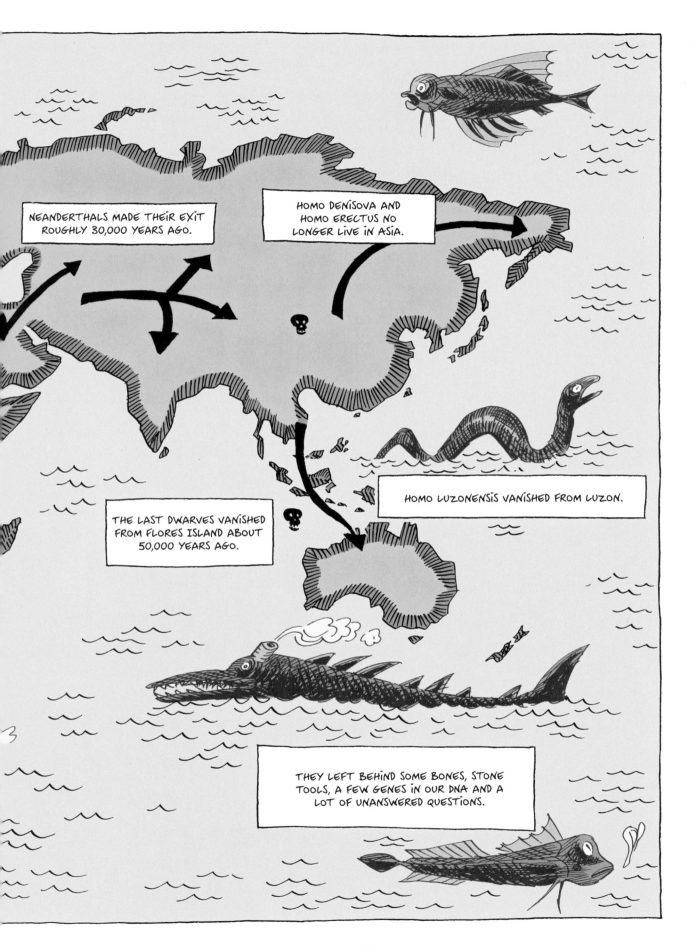

NEANDERTHALS MADE THEIR EXIT ROUGHLY 30,000 YEARS AGO.

HOMO DENISOVA AND HOMO ERECTUS NO LONGER LIVE IN ASIA.

THE LAST DWARVES VANISHED FROM FLORES ISLAND ABOUT 50,000 YEARS AGO.

HOMO LUZONENSIS VANISHED FROM LUZON.

THEY LEFT BEHIND SOME BONES, STONE TOOLS, A FEW GENES IN OUR DNA AND A LOT OF UNANSWERED QUESTIONS.

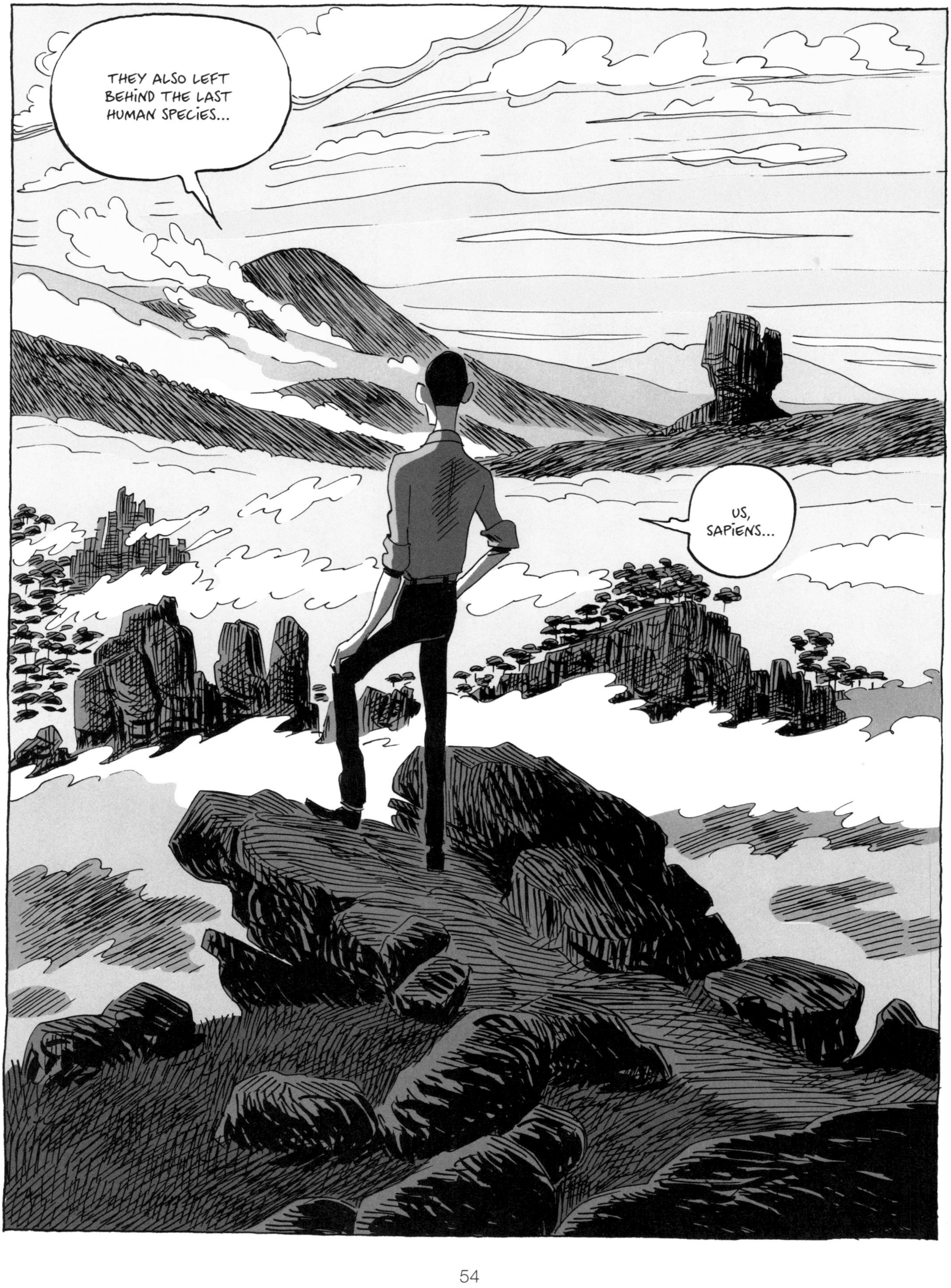
THEY ALSO LEFT BEHIND THE LAST HUMAN SPECIES...
US, SAPIENS...

COMING SOON iN SAPiENS

WHAT WAS SAPIENS' SECRET TO SUCCESS?

SUPER-SAPIENS

HOW DID SAPIENS MANAGE TO SETTLE SO RAPIDLY IN SO MANY DIFFERENT FAR-FLUNG HABITATS?

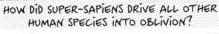

HOW DID SUPER-SAPIENS DRIVE ALL OTHER HUMAN SPECIES INTO OBLIVION?

HOW COME EVEN THE MUSCULAR NEANDERTHALS COULDN'T SURVIVE THEIR INVASION?

HOW DID SUPER-SAPIENS SUCCEED IN CONQUERING THE WORLD?

FIND THE ANSWERS TO ALL THESE QUESTIONS IN THE NEXT ADVENTURES OF **SUPER-SAPIENS!**

MASTERS OF FICTION

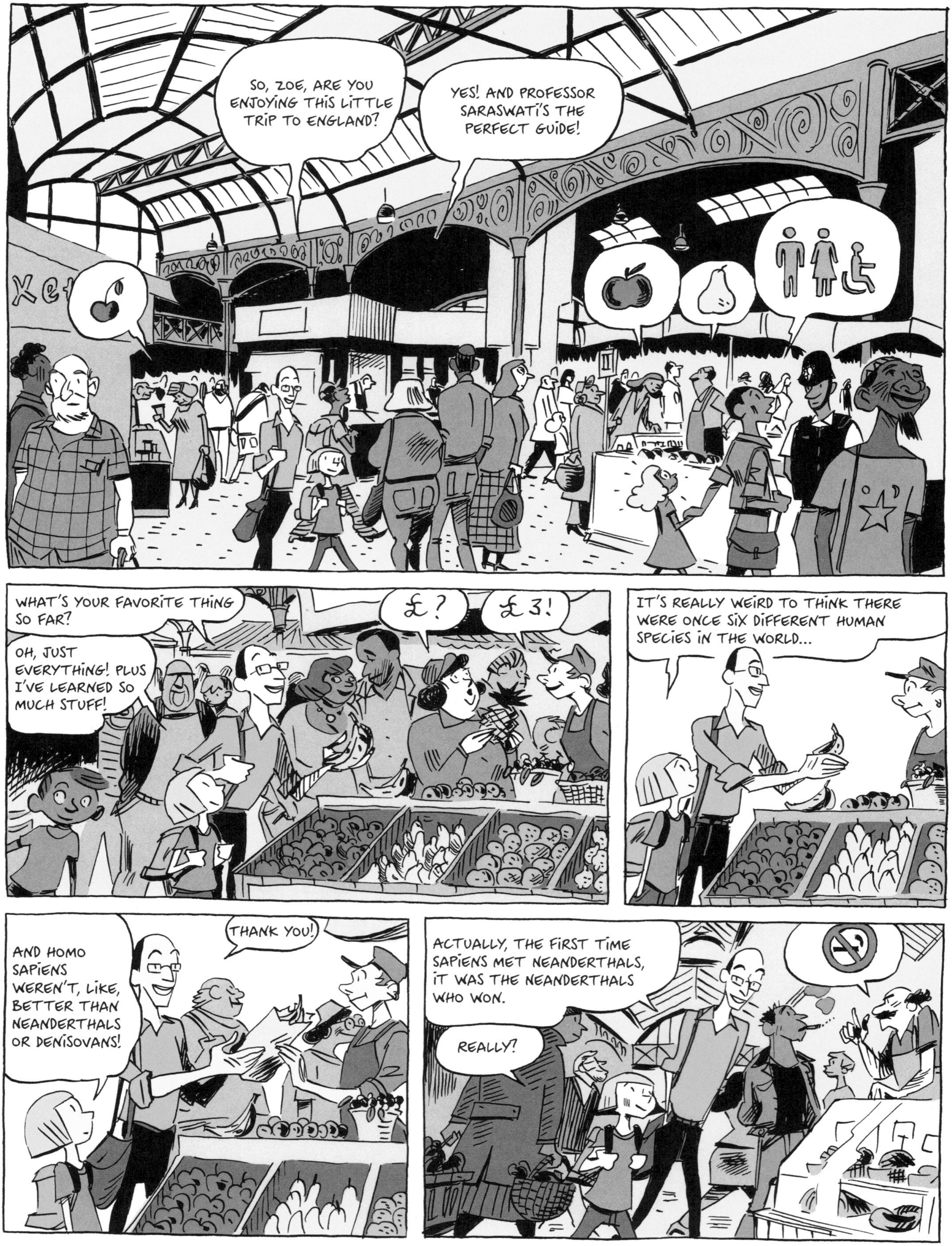
SO, ZOE, ARE YOU ENJOYING THIS LITTLE TRIP TO ENGLAND?
YES! AND PROFESSOR SARASWATI'S THE PERFECT GUIDE!
WHAT'S YOUR FAVORITE THING SO FAR?
OH, JUST EVERYTHING! PLUS I'VE LEARNED SO MUCH STUFF!
£?
£3!
IT'S REALLY WEIRD TO THINK THERE WERE ONCE SIX DIFFERENT HUMAN SPECIES IN THE WORLD...
AND HOMO SAPIENS WEREN'T, LIKE, BETTER THAN NEANDERTHALS OR DENISOVANS!
THANK YOU!
ACTUALLY, THE FIRST TIME SAPIENS MET NEANDERTHALS, IT WAS THE NEANDERTHALS WHO WON.
REALLY?

SJJ483D698YZHDQ64QIDGDJKSDDJ995SF

ABOUT 100,000 YEARS AGO, SOME GROUPS OF SAPIENS VENTURED INTO NEANDERTHAL TERRITORY IN THE MIDDLE EAST, BUT THEY DIDN'T STAY THERE LONG.

BECAUSE THE NEANDERTHALS WERE TOO STRONG?

PERHAPS... OR MAYBE THE CLIMATE WAS TOO COLD FOR THEM. OR MAYBE THEY HAD NO IMMUNITY TO THE LOCAL PARASITES...

SJJ483D6...

WE DON'T KNOW FOR SURE. WHATEVER THE REASON, THE SAPIENS EVENTUALLY RETREATED, LEAVING THE MIDDLE EAST TO THE NEANDERTHALS.

RETREAT! RETREAT!

IT'S NOT A GREAT RECORD OF ACHIEVEMENT! WHICH IS WHY SOME SCHOLARS SPECULATE THAT THE SAPIENS BACK THEN WERE STILL VERY DIFFERENT FROM US. THEY LOOKED LIKE US, AND THEIR BRAINS WERE AS BIG AS OURS, BUT THE INTERNAL STRUCTURE OF THEIR BRAINS WAS PROBABLY DIFFERENT.

THEIR COGNITIVE ABILITIES—THINGS LIKE LEARNING, REMEMBERING AND COMMUNICATING— WERE FAR MORE LIMITED.

YOU PROBABLY COULDN'T TEACH THEM TO SPEAK ENGLISH, NOR EXPLAIN CHRISTIAN DOGMA OR THE THEORY OF EVOLUTION.

DDJ995SF

THE REVENGE OF THE SAPIENS

...ALSO, THE FIRST OBJECTS THAT WE CAN RELIABLY CALL ART OR JEWELRY. AND THE FIRST CLEAR EVIDENCE FOR RELIGION, TRADING AND COMPLEX SOCIAL ORDER.

An ivory figurine of a "lion-man" (or "lioness-woman") found in a cave in Stadel, Germany. Approximately 32,000 years old.

WHAT EXACTLY HAPPENED 70,000 YEARS AGO THAT MADE SAPIENS START DOING ALL THIS?

MOST RESEARCHERS SAY IT CAME DOWN TO A REVOLUTION IN SAPIENS' COGNITIVE ABILITIES.

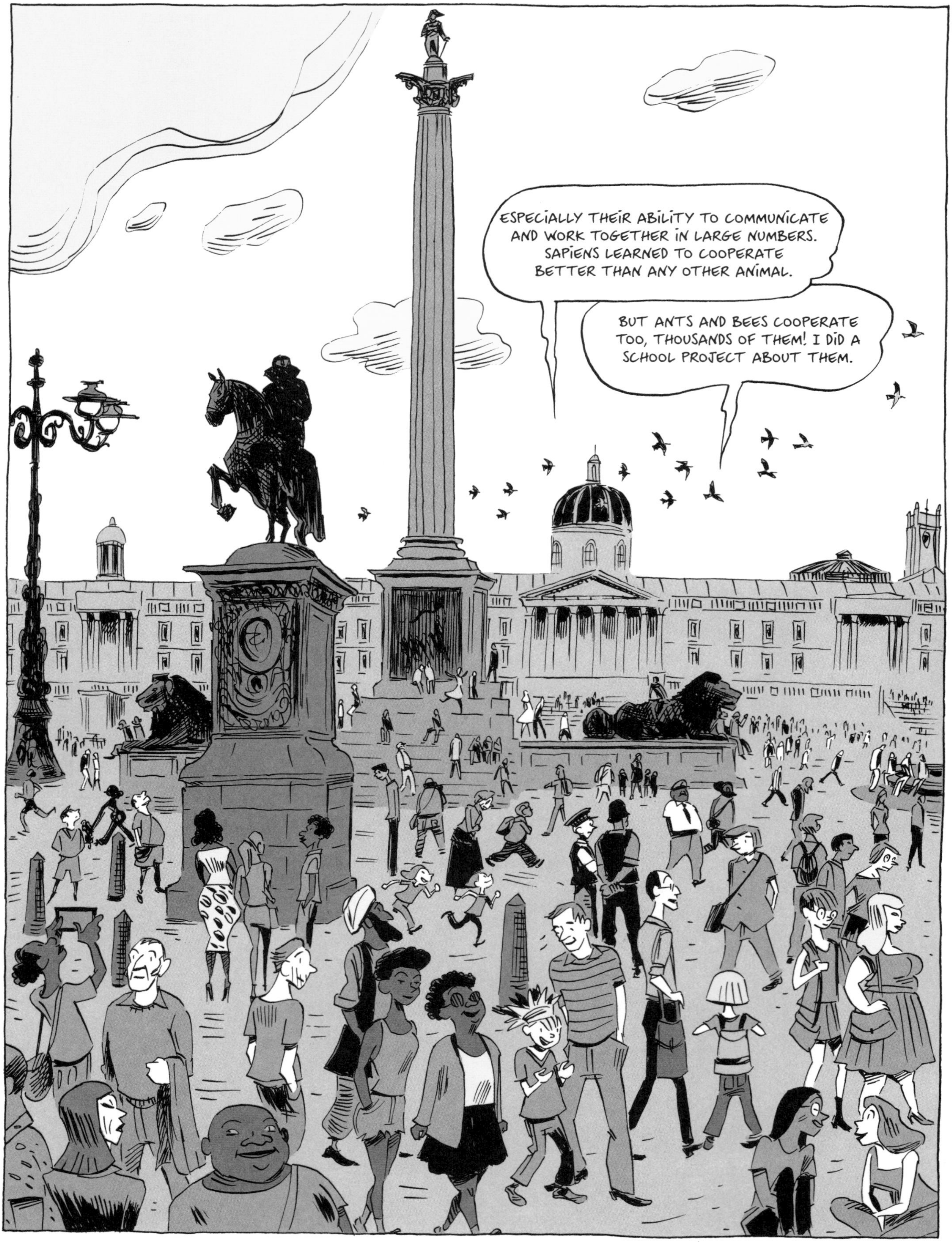
ESPECIALLY THEIR ABILITY TO COMMUNICATE AND WORK TOGETHER IN LARGE NUMBERS. SAPIENS LEARNED TO COOPERATE BETTER THAN ANY OTHER ANIMAL.
BUT ANTS AND BEES COOPERATE TOO, THOUSANDS OF THEM! I DID A SCHOOL PROJECT ABOUT THEM.

YOU'RE RIGHT! BUT THEY ONLY COOPERATE WITH CLOSE RELATIVES, AND THEY'RE NOT VERY FLEXIBLE.

IF A BEEHIVE FACES A NEW THREAT OR A NEW OPPORTUNITY, BEES CAN'T REINVENT THE HIVE'S SOCIAL SYSTEM TO HELP DEAL WITH IT.

FOR EXAMPLE, THEY CAN'T EXECUTE THEIR QUEEN AND ESTABLISH A REPUBLIC.

WOLVES AND CHIMPANZEES COOPERATE FAR MORE FLEXIBLY THAN BEES, BUT ONLY IN SMALL GROUPS.

BECAUSE COOPERATION AMONG CHIMPANZEES DEPENDS ON CLOSE PERSONAL RELATIONS.

CHIMPS DON'T COOPERATE WITH STRANGERS.

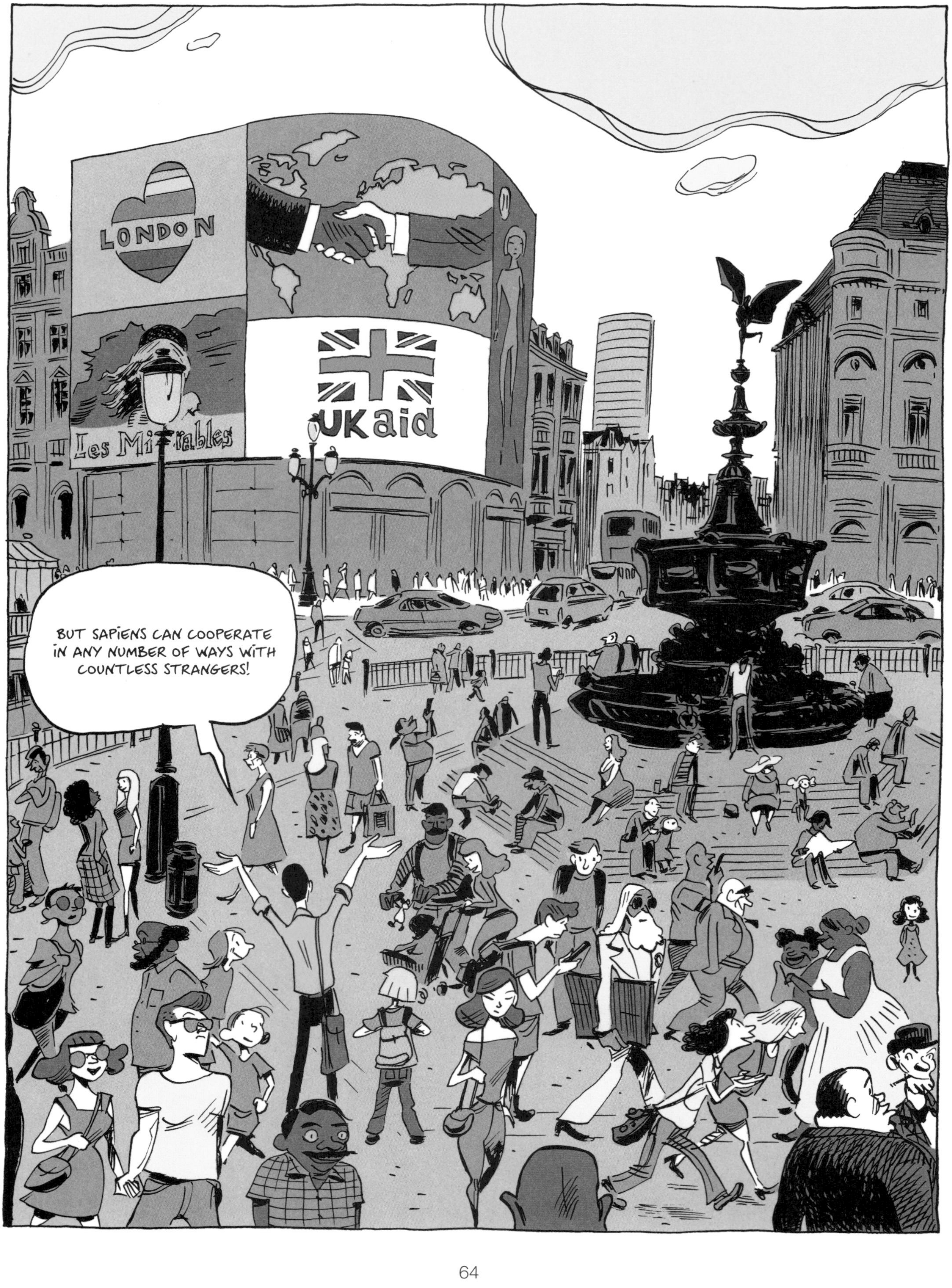
LONDON
Les Misérables
UKaid
BUT SAPIENS CAN COOPERATE IN ANY NUMBER OF WAYS WITH COUNTLESS STRANGERS!

THAT'S WHY SAPIENS RULE THE WORLD, WHILE ANTS EAT OUR LEFTOVERS AND POOR CHIMPS ARE LOCKED UP IN ZOOS AND RESEARCH LABORATORIES...

THAT REMINDS ME, A FRIEND SENT ME A LINK TO A REALLY INTERESTING CLIP ABOUT CHIMPS. BUT I HAVEN'T SEEN IT YET.

GREAT! WHERE'S YOUR TABLET?

I THINK THIS IS IT...

THE EXTRAORDINARY LIFE OF CHIMPANZEES

OUR CHIMPANZEE COUSINS USUALLY LIVE IN SMALL TROOPS OF SEVERAL DOZEN INDIVIDUALS. THEY FORM CLOSE FRIENDSHIPS, HUNT TOGETHER AND FIGHT SHOULDER-TO-SHOULDER AGAINST BABOONS, CHEETAHS AND ENEMY CHIMPANZEES.

THEIR SOCIAL STRUCTURE TENDS TO BE HIERARCHICAL. THE DOMINANT MEMBER, ALMOST ALWAYS A MALE, IS TERMED THE "ALPHA MALE." THE OTHER CHIMPS SHOW SUBMISSION TO THE ALPHA MALE BY BOWING BEFORE HIM.

NOT UNLIKE HUMAN SUBJECTS KNEELING BEFORE A KING.

THE ALPHA MALE TRIES TO MAINTAIN SOCIAL HARMONY WITHIN HIS TROOP. WHEN TWO INDIVIDUALS FIGHT, HE INTERVENES AND STOPS THE VIOLENCE.

HE CAN BE SELFISH TOO. HE MIGHT MONOPOLIZE PARTICULARLY PRIZED FOODS AND PREVENT LOWER-RANKING MALES FROM MATING WITH THE FEMALES.

WHEN TWO MALES ARE CONTESTING THE ALPHA POSITION, THEY USUALLY ESTABLISH EXTENSIVE COALITIONS OF SUPPORTERS, BOTH MALE AND FEMALE. TIES BETWEEN COALITION MEMBERS ARE BASED ON INTIMATE DAILY CONTACT—HUGGING, TOUCHING, KISSING, GROOMING AND MUTUAL FAVORS.

A BIT LIKE HUMAN POLITICIANS ON ELECTION CAMPAIGNS GOING AROUND SHAKING HANDS AND KISSING BABIES...

...CONTENDERS FOR THE TOP POSITION IN A CHIMPANZEE GROUP DO A LOT OF HUGGING, BACK-SLAPPING AND KISSING BABY CHIMPS. THE ALPHA MALE USUALLY WINS HIS POSITION NOT BECAUSE HE'S PHYSICALLY STRONGER, BUT BECAUSE HE LEADS A LARGE STABLE COALITION.

THESE COALITIONS ALSO PLAY A CENTRAL PART IN DAY-TO-DAY ACTIVITIES. MEMBERS OF A COALITION SPEND MORE TIME TOGETHER, SHARE FOOD AND HELP ONE ANOTHER.

THERE ARE CLEAR LIMITS TO THE SIZE OF THESE GROUPS. FOR A GROUP TO FUNCTION, ALL ITS MEMBERS MUST KNOW EACH OTHER INTIMATELY. TWO CHIMPANZEES WHO'VE NEVER MET WON'T KNOW WHETHER THEY CAN TRUST ONE ANOTHER, WHETHER IT'S WORTH HELPING ONE ANOTHER AND WHICH OF THEM RANKS HIGHER.

UNDER NATURAL CONDITIONS, A TYPICAL CHIMPANZEE TROOP NUMBERS PERHAPS 20 TO 50 INDIVIDUALS. AS A TROOP GETS BIGGER, THE SOCIAL ORDER DESTABILIZES, EVENTUALLY LEADING TO A SPLIT WITH SOME OF THE ANIMALS FORMING A BREAKAWAY TROOP.

ZOOLOGISTS HAVE SEEN ONLY A HANDFUL OF GROUPS LARGER THAN 100. SEPARATE GROUPS SELDOM COOPERATE, AND TEND TO COMPETE FOR TERRITORY AND FOOD. RESEARCHERS HAVE DOCUMENTED PROLONGED WARFARE BETWEEN GROUPS, AND EVEN ONE CASE OF "GENOCIDAL" ACTIVITY WHEN ONE TROOP SYSTEMATICALLY SLAUGHTERED MOST MEMBERS OF A NEIGHBORING BAND.

ANCIENT HUMANS WERE MUCH LIKE CHIMPANZEES. EVEN TODAY, WE SAPIENS ARE STILL EMBARRASSINGLY SIMILAR TO CHIMPANZEES—AS INDIVIDUALS AND IN FAMILIES.

BUT IT'S POINTLESS LOOKING FOR DIFFERENCES BETWEEN SAPIENS AND OTHER ANIMALS AT THE LEVEL OF INDIVIDUALS OR FAMILIES.

WHAT MAKES US REALLY DIFFERENT IS THE WAY WE COOPERATE IN LARGE NUMBERS.

IMAGINE THE PANDEMONIUM IF YOU TRIED TO BUNCH TOGETHER THOUSANDS OF CHIMPANZEES IN MARACANA STADIUM, WALL STREET, THE VATICAN OR THE BRITISH HOUSES OF PARLIAMENT...

AND YET SAPIENS REGULARLY GATHER IN THEIR THOUSANDS IN PLACES LIKE THESE. TOGETHER, THEY CAN SET UP ORDERLY SYSTEMS OF COOPERATION SUCH AS TRADE NETWORKS, MASS CELEBRATIONS AND POLITICAL INSTITUTIONS.

THE REAL DIFFERENCE BETWEEN US AND CHIMPANZEES IS THE MYSTERIOUS GLUE THAT BINDS LARGE NUMBERS OF INDIVIDUALS, FAMILIES AND GROUPS. THIS GLUE MADE US THE MASTERS OF THE WORLD.

OF COURSE, WE ALSO NEEDED OTHER SKILLS, LIKE MAKING AND USING TOOLS. BUT EVEN TOOLMAKING WAS RELATIVELY INCONSEQUENTIAL UNTIL IT WAS COUPLED WITH OUR ABILITY FOR MASS COOPERATION.

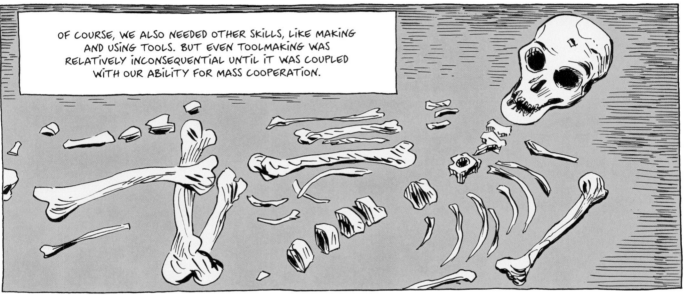

HOW COME TODAY WE HAVE THIS....

SATAN 2 MISSILE. CAPABLE OF DESTROYING NEW YORK OR MOSCOW.

...WHEN 30,000 YEARS AGO, ALL WE HAD WAS THIS?

SPEAR WITH STONE POINT. CAPABLE OF KILLING A ZEBRA.

PHYSIOLOGICALLY, OUR TOOL-MAKING CAPACITY HASN'T IMPROVED SIGNIFICANTLY IN THESE 30,000 YEARS.

NOBEL PRIZE

ALBERT EINSTEIN WAS NOWHERE NEAR AS GOOD WITH HIS HANDS AS AN ANCIENT HUNTER-GATHERER.

OUCH!

NOBEL PRIZE

BUT OUR CAPACITY TO COOPERATE WITH LARGE NUMBERS OF STRANGERS HAS IMPROVED DRAMATICALLY. A SINGLE PERSON COULD MAKE AN ANCIENT FLINT SPEARHEAD IN MINUTES WITH ADVICE AND HELP FROM A FEW INTIMATE FRIENDS.

PRODUCING A MODERN NUCLEAR WARHEAD REQUIRES THE COOPERATION OF MILLIONS OF STRANGERS ALL OVER THE WORLD—FROM PEOPLE MINING THE URANIUM ORE DEEP UNDERGROUND TO THEORETICAL PHYSICISTS WRITING LONG MATHEMATICAL FORMULAS ABOUT THE INTERACTIONS OF SUBATOMIC PARTICLES.

ALL OF HOMO SAPIENS' GREAT ACHIEVEMENTS—FROM BUILDING THE PYRAMIDS TO LANDING ON THE MOON—ARE BASED ON LARGE-SCALE COOPERATION.

THAT MAKES SENSE.

BUT HOW COME SAPIENS CAN COOPERATE IN SUCH BIG NUMBERS?

IT'S ALL DOWN TO THOSE NEW COMMUNICATION SKILLS THAT SAPIENS STARTED USING ABOUT 70,000 YEARS AGO.

OKAY, SO WHAT ARE THESE SUPER-SPECIAL COMMUNICATION SKILLS?

IF YOU LIKE, WE COULD PAY A VISIT TO PROFESSOR ROBIN DUNBAR...

UNDERGROUND

HE'S AN EXPERT ON HUMAN COMMUNICATION.

LATER, AT PROFESSOR DUNBAR'S HOME....

EXCELLENT IDEA TO COME AND VISIT ME WITH YOUR NIECE, MY DEAR YUVAL!

WOULD YOU LIKE A FRUIT JUICE, ZOE?

YES PLEASE, PROFESSOR.

A GLASS OF WINE, YUVAL?

NOT FOR ME THANKS. FRUIT JUICE IS PERFECT FOR ME TOO.

I'LL HAVE A LITTLE GRAPE JUICE MYSELF, HA! HA!

HAVE YOU HEARD THE LATEST SCANDAL IN THE PSYCHOLOGY DEPARTMENT?

NO, CAN'T SAY I HAVE...

REALLY?! YOU WON'T BELIEVE THIS! LESLIE AND BRANDON HAVE SPLIT UP! SHE CAUGHT HIM IN THE PUB WITH A GIRL YOUNG ENOUGH TO BE HIS DAUGHTER! IT'S ALL ANYONE CAN TALK ABOUT IN THE CAFETERIA!

SQUAWK!

AHA! I'M SURE IT'S A FASCINATING STORY, PROFESSOR...

BUT ZOE AND I DON'T HAVE TIME TO CHAT FOR LONG. WE ACTUALLY CAME TO ASK YOU ABOUT COMMUNICATION.

WELL, I'M JUST AS HAPPY TO TALK ABOUT COMMUNICATION! THAT AND A NICE GLASS OF WINE—NOTHING LIKE IT!

WHAT I'D LIKE TO KNOW IS WHAT MAKES SAPIENS' COMMUNICATION SKILLS SO SPECIAL?

I SEE, YES, THAT'S A VERY GOOD QUESTION.

WHERE TO START...

FIRST OF ALL, I SHOULD POINT OUT THAT COMMUNICATION IS NOT UNIQUE TO SAPIENS. EVEN INSECTS COMMUNICATE WITH EACH OTHER.

ANTS TELL EACH OTHER WHERE THEY CAN FIND FOOD BY EXCHANGING CHEMICALS.

BEES TRANSMIT INFORMATION BY DANCING...

I CAN SHOW YOU A GOOD EXAMPLE.

LOTS OF ANIMALS, INCLUDING ALL APE AND MONKEY SPECIES, USE VOCAL SIGNALS. ZOOLOGISTS HAVE STUDIED THE DIFFERENT SIGNALS THAT GREEN MONKEYS COMMUNICATE WITH.

KYAK! KYAK! KYAK! KYAK!*

*WATCH OUT, A LION!

Yi-AAKH! Yi-AAKH *

* WATCH OUT! AN EAGLE!

LOOK WHAT HAPPENED WHEN RESEARCHERS PLAYED RECORDINGS OF THESE TWO WARNINGS TO A GROUP OF MONKEYS...

Yi-AAKH! Yi-AAKH*

KYAK! KYAK! KYAK!

SAPIENS CAN PRODUCE MANY MORE DISTINCT SOUNDS THAN GREEN MONKEYS, BUT IT ISN'T THESE VOCAL SKILLS THEMSELVES THAT MAKE US SO SPECIAL.

MY PARROTS CAN MIMIC ALL THE SOUNDS I MAKE, AND PLENTY OF OTHERS THAT I CAN'T!

MIMIC ALL THE SOUNDZZZZ!

DING DONG!

JUST A MINUTE, THERE'S SOMEONE AT THE DOOR.

HA! HA!

BUT THEN WHAT'S SO SPECIAL ABOUT HOW WE SAPIENS COMMUNICATE?

FIRST OF ALL,

OUR LANGUAGE

IS AMAZINGLY

SUPPLE.

WE CAN CONNECT A LIMITED NUMBER OF SOUNDS AND SIGNS TO PRODUCE AN INFINITE NUMBER OF SENTENCES, EACH WITH A DISTINCT MEANING.

THAT'S HOW WE ASSIMILATE, STORE AND COMMUNICATE A HUGE AMOUNT OF INFORMATION ABOUT THE WORLD AROUND US.

A GREEN MONKEY CAN WARN, "CAREFUL! A LION!" BUT A SAPIENS CAN DO MUCH MORE THAN THAT...

I WAS OUT GATHERING NUTS AS USUAL THIS MORNING WHEN I SAW A BIG LION FOLLOWING A HERD OF BISON. THE BISON WERE WALKING ALONG THE RIVER, HEADING FOR THE CROSSING. BUT THEY WERE MOVING VERY SLOWLY, BECAUSE OF THE LION.

IF WE'RE QUICK, WE CAN DRIVE AWAY THE LION AND HUNT THE BISON BEFORE THEY CROSS THE RIVER.

GREAT IDEA! LET'S TELL EVERYONE ELSE.

NOW THAT THEY HAVE THE INFORMATION, ALL THE MEMBERS OF THE GROUP CAN THINK TOGETHER AND MAKE A PLAN.

WE CAN SPLIT INTO THREE GROUPS, SURROUND THE BISON AND TRAP THEM BY THE RIVER.

WE'LL RUN TO THE CROSSING PLACE TO BLOCK IT OFF.

WE'LL COME AT THEM FROM THE SOUTH.

WE'LL CREEP THROUGH THE THICKET.

EXCHANGING INFORMATION ABOUT LIONS, BISON AND RIVERS WAS VERY USEFUL...

BUT THE MOST IMPORTANT INFORMATION HUMANS EXCHANGED WASN'T ABOUT LIONS OR BISON—IT WAS ABOUT HUMANS THEMSELVES.

BASICALLY, HOMO SAPIENS IS A SOCIAL ANIMAL.

CHEERS!

SOCIAL COOPERATION IS OUR KEY FOR SURVIVAL AND REPRODUCTION.

WHETHER WE'RE HUNTING BISON OR BUILDING ATOM BOMBS, WE NEED TO TRUST ONE ANOTHER.

SO IT'S NOT ENOUGH JUST KNOWING WHERE TO FIND BISON OR URANIUM MINES?

EXACTLY RIGHT!

IT'S MUCH MORE IMPORTANT FOR PEOPLE TO KNOW WHO IN THEIR IMMEDIATE CIRCLE HATES WHO, WHO'S SLEEPING WITH WHO, WHO'S HONEST, WHO'S A CHEAT...

OH!

WHAT'S GOING ON?

BREAKING NEWS

LIVE

OH, MY GOODNESS!

PFFF, OLD NEWS. IT WAS ALL OVER TWITTER LAST NIGHT!

BRAD AND ANGELINA: THE BREAK-UP?

SORRY, THAT PUT ME OFF TRACK... WHERE WAS I?

OH YES!

IF YOU WANT TO KEEP TABS ON THE EVER-CHANGING RELATIONSHIPS OF EVEN A FEW DOZEN PEOPLE YOU NEED TO OBTAIN AND PROCESS A STAGGERING AMOUNT OF INFORMATION.

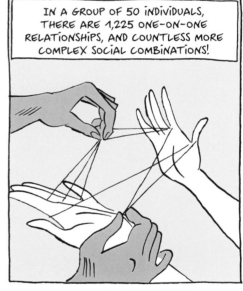

IN A GROUP OF 50 INDIVIDUALS, THERE ARE 1,225 ONE-ON-ONE RELATIONSHIPS, AND COUNTLESS MORE COMPLEX SOCIAL COMBINATIONS!

CHIMPS STRUGGLE TO MAINTAIN COHESION IN A BAND WITH 50 INDIVIDUALS.

BECAUSE THEY CAN'T KEEP UP WITH THE LATEST GOSSIP?

EXACTLY! IF TWO CHIMP FRIENDS HAVE A PUNCH-UP, HOW WILL THE OTHERS KNOW ABOUT IT IF THEY HAVEN'T SEEN IT WITH THEIR OWN EYES? CHIMPANZEES ARE ALWAYS HUNGRY FOR SOCIAL INFORMATION, BUT THEY'RE NOT VERY GOOD AT GOSSIPING.

IT'S MUCH EASIER FOR SAPIENS BECAUSE WE SPEND SO MUCH TIME TALKING BEHIND EACH OTHER'S BACKS!

BRAD PITT WAS ALLEGEDLY SPOTTED IN TEARS, WANDERING AROUND HOLLYWOOD BOULEVARD ALONE...

PAH... BUT WHAT A STORY!

BUT SURELY IT'S WRONG TO TALK ABOUT PEOPLE BEHIND THEIR BACKS?

IT'S A MUCH-MALIGNED ACTIVITY, BUT IT'S ACTUALLY ESSENTIAL FOR COOPERATION IN LARGE NUMBERS.

WHEN SAPIENS ACQUIRED NEW LINGUISTIC SKILLS ABOUT 70,000 YEARS AGO THEY COULD GOSSIP FOR HOURS ON END.

RELIABLE INFORMATION ABOUT WHO COULD BE TRUSTED MEANT THAT SMALL BANDS COULD EXPAND INTO LARGER BANDS, AND SAPIENS COULD DEVELOP TIGHTER AND MORE SOPHISTICATED TYPES OF COOPERATION.

NEANDERTHAL →

← SAPIENS

IT MIGHT SOUND LIKE A JOKE, BUT THERE ARE LOTS OF STUDIES THAT SUPPORT PROFESSOR DUNBAR'S GOSSIP THEORY.

PING!

SORRY, THAT WAS MY PHONE.

PLEASE, GO AHEAD, ZOE.

EVEN TODAY, MOST HUMAN COMMUNICATION IS STILL GOSSIP.

HI ZO, U HERD ABT BRANGELINA?

KELLY SEZ ITS BRAD'S FAULT.

IM TELG U HE WOZ WTH LINDA, OBVS.

GOSSIP COMES SO NATURALLY TO US IT'S AS IF THAT'S WHAT OUR LANGUAGE EVOLVED FOR IN THE FIRST PLACE.

DO YOU REALLY THINK THAT WHEN HISTORY PROFESSORS MEET FOR LUNCH THEY TALK ABOUT WHAT SPARKED WORLD WAR I?

OR NUCLEAR PHYSICISTS SPEND THEIR COFFEE BREAKS AT SCIENTIFIC CONFERENCES TALKING ABOUT QUARKS?

THEY'RE FAR MORE LIKELY TO GOSSIP ABOUT THE PROFESSOR WHO CAUGHT HER HUSBAND CHEATING, OR THE QUARREL BETWEEN THE HEAD OF THE DEPARTMENT AND THE DEAN.

OR RUMORS THAT A COLLEAGUE USED HIS RESEARCH FUNDS TO BUY A LEXUS.

GOSSIP USUALLY FOCUSES ON WRONGDOINGS, SO IT HELPS ENFORCE SOCIAL NORMS AND MAINTAIN GROUP COHESION

INSPIRED BY NORMAN ROCKWELL

BUT THERE'S STILL A PROBLEM BECAUSE EVEN GOSSIP HAS ITS LIMITS. NO MATTER HOW MUCH YOU GOSSIP, MOST PEOPLE CAN'T BE REALLY CLOSE TO MORE THAN 150 OTHER PEOPLE.

5 CLOSE FRIENDS

500 ACQUAINTANCES

THAT'S WHY THERE'S A CRITICAL THRESHOLD IN HUMAN ORGANIZATIONS SOMEWHERE AROUND THIS MAGIC NUMBER—150.

1,500 PEOPLE YOU WOULD RECOGNIZE IN THE STREET

BELOW THE 150 THRESHOLD, COMMUNITIES, BUSINESSES, SOCIAL NETWORKS AND MILITARY UNITS CAN KEEP GOING MOSTLY THANKS TO INTIMATE ACQUAINTANCE AND RUMOR-MONGERING. THERE'S NO NEED FOR FORMAL RANKS, TITLES AND LAW BOOKS TO KEEP ORDER.

A PLATOON OF 30 SOLDIERS OR EVEN A COMPANY OF 100 SOLDIERS CAN FUNCTION WELL ON THE BASIS OF CLOSE TIES, WITH A MINIMUM OF FORMAL DISCIPLINE.

A WELL-RESPECTED SERGEANT CAN BECOME THE UNOFFICIAL "KING OF THE COMPANY" AND CAN EVEN TELL OFFICERS WHAT TO DO.

BUT ONCE YOU CROSS THAT THRESHOLD OF 150 INDIVIDUALS, THINGS CAN'T WORK LIKE THAT.

PASSES! I TOLD YOU I WANT TO SEE PASSES!

LET'S GO UP ON THE ROOF, IT'S THE PERFECT PLACE FOR THIS LAST POINT.

WHAT A WEIRD BUILDING, IT LOOKS LIKE A WALKIE-TALKIE!

YES, AND THAT'S EXACTLY WHAT WE CALL IT.

YOU CAN'T RUN A DIVISION WITH THOUSANDS OF SOLDIERS THE SAME WAY YOU RUN A PLATOON. AND SUCCESSFUL FAMILY BUSINESSES USUALLY FACE A CRISIS WHEN THEY GROW LARGER AND HIRE MORE PERSONNEL. IF THEY CAN'T REINVENT THEMSELVES, THEY GO BUST.

HAVE A LISTEN, IT'S FUN.

HELLO, MR. JONES, I'M EXPECTING THE PLUMBER WHILE I'M OUT, CAN I LEAVE MY KEY WITH YOU?

OF COURSE, MRS. PEPPERPOT!

WHO THE HELL ARE YOU?

I'M THE MARKETING MANAGER.

SINCE WHEN?

ABOUT TWO YEARS. AND YOU, MADAM?

I'M YOUR CEO.

PROFESSOR, HOW DO YOU THINK HOMO SAPIENS MANAGED TO CROSS THE 150 THRESHOLD? HOW DID WE END UP FOUNDING CITIES WITH TENS OF THOUSANDS OF INHABITANTS AND EMPIRES RULING HUNDREDS OF MILLIONS?

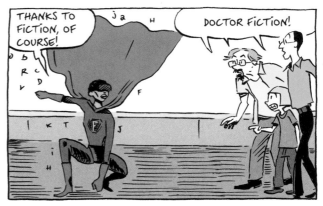

THANKS TO FICTION, OF COURSE!

DOCTOR FICTION!

THANKS, PROF! YOU WERE GREAT! I'LL TAKE THIS FROM HERE!

BUT.... WELL, I JUST WANTED TO SAY...

SO.... WHAT'S THE SECRET?

THE SECRET OF SUCCESS IS MYTHOLOGY, RIGHT?

LARGE NUMBERS OF TOTAL STRANGERS CAN COOPERATE SUCCESSFULLY IF THEY BELIEVE IN THE SAME MYTHS!

SAPIENS RULE THE WORLD BECAUSE THEY'RE THE ONLY ANIMALS CAPABLE OF CREATING AND BELIEVING FICTIONAL STORIES.

AND AS LONG AS EVERYONE BELIEVES IN THE SAME FICTIONS, EVERYONE FOLLOWS THE SAME RULES.

LET ME GUESS... I BET THE PROFESSOR SHOWED YOU HOW ANIMALS USE SIGNALS TO DESCRIBE ALL KINDS OF STUFF. AM I RIGHT?

HE DID, YES. A CHIMPANZEE CAN SAY: "WATCH OUT! THERE'S A LION!" OR "LOOK, THERE'S A BANANA, LET'S GET IT."

THAT'S RIGHT! BUT WE SAPIENS DON'T USE LANGUAGE ONLY TO DESCRIBE THINGS THAT WE SEE AROUND US—WE CAN ALSO USE LANGUAGE TO INVENT STUFF. FICTIONS!

A SAPIENS CAN SAY "LOOK UP THERE! THERE'S A GOD ABOVE THE CLOUDS AND HE'LL PUNISH YOU IF YOU DON'T DO AS I SAY."

AND IF YOU ALL BELIEVE IN THAT STORY, THEN YOU'LL ALL FOLLOW THE SAME LAWS AND RULES, SO YOU'LL BE ABLE TO COOPERATE EFFECTIVELY, EVEN IF YOU DON'T KNOW EACH OTHER.

THAT'S SOMETHING ONLY SAPIENS CAN DO.

YOU COULD NEVER CONVINCE A CHIMPANZEE TO GIVE YOU A BANANA BY PROMISING HIM UNLIMITED BANANAS IN APE HEAVEN.

NO CHIMP WOULD EVER BELIEVE A STORY LIKE THAT! ONLY SAPIENS MIGHT BELIEVE IT. AND THAT'S HOW WE CAN COOPERATE WITH MILLIONS OF STRANGERS, WHEREAS SKEPTICAL CHIMPS CAN'T.

ALL LARGE-SCALE HUMAN COOPERATION DEPENDS ON COMMON MYTHS THAT EXIST ONLY IN PEOPLE'S COLLECTIVE IMAGINATION. THAT'S TRUE OF A PREHISTORIC TRIBE...

...AN ANCIENT CITY...

...A MEDIEVAL CHURCH...

...OR A MODERN STATE...

CHURCHES ARE BASED ON COMMON RELIGIOUS MYTHS.

TWO CATHOLICS WHO'VE NEVER MET MIGHT GO ON CRUSADE TOGETHER OR POOL FUNDS TO BUILD A HOSPITAL BECAUSE THEY BOTH BELIEVE THAT GOD WAS INCARNATED IN HUMAN FLESH AND ALLOWED HIMSELF TO BE CRUCIFIED TO REDEEM OUR SINS.

STATES ARE ROOTED IN COMMON NATIONAL MYTHS.

A WELSHMAN AND AN ENGLISHMAN WHO'VE NEVER MET MIGHT RISK THEIR LIVES TO SAVE EACH OTHER BECAUSE THEY BOTH BELIEVE IN GREAT BRITAIN.

JUDICIAL SYSTEMS ARE ROOTED IN COMMON LEGAL MYTHS.

TWO LAWYERS WHO'VE NEVER MET MIGHT WORK TOGETHER TO DEFEND A COMPLETE STRANGER BECAUSE THEY BOTH BELIEVE IN LAWS, JUSTICE, HUMAN RIGHTS—AND THE MONEY PAID OUT IN FEES.

YET NONE OF THESE THINGS EXISTS OUTSIDE THE STORIES THAT PEOPLE INVENT AND TELL ONE ANOTHER.

CAPE CANAVERAL TO EXPLORER 412. SO? HAVE YOU FOUND ANY GODS, NATIONS OR HUMAN RIGHTS?

EXPLORER 412 TO CAPE CANAVERAL: NOT REALLY....

WHAT ABOUT JUSTICE AND MONEY!

NO.... BUT WE HAVE FOUND A WHOLE LOT OF HYDROGEN....

PEOPLE EASILY GET THE IDEA THAT "PRIMITIVE TRIBES" CEMENT THEIR SOCIAL ORDER BY BELIEVING IN IMAGINARY SPIRITS.

WHAT WE DON'T REALIZE IS THAT OUR MODERN INSTITUTIONS FUNCTION IN EXACTLY THE SAME WAY!

STOP CUTTING DOWN TREES. YOU'LL ANGER THE SPIRITS OF THE FOREST!

OF COURSE, OF COURSE...

BUT WE MUSTN'T ANGER THE SPIRITS OF THE STOCK MARKET EITHER, I'M SURE YOU UNDERSTAND...

MODERN FIRMS AREN'T SO VERY DIFFERENT FROM ANCIENT TRIBES.

MODERN BUSINESS-PEOPLE AND LAWYERS ARE, IN FACT, POWERFUL SORCERERS!

THE BIG DIFFERENCE BETWEEN THEM AND TRIBAL SHAMANS IS THAT MODERN LAWYERS TELL WAY WEIRDER STORIES!

I CAN EXPLAIN ALL THIS WITH A NICE EXAMPLE.

LOOK AT MY CAR... IT'S A PEUGEOT.

YOU CAN TELL FROM ITS LOGO, HERE.

REMIND YOU OF ANYTHING, HUH?

WAIT, I'LL REFRESH YOUR MEMORY...

HEY! HELLO THERE, DOCTOR FICTION!

HELLO, BILL!

WHERE ARE YOU HEADED?

I'M DRIVING BACK TO THE CAVE. THERE'S A WAY TO GO BEFORE I REACH STADEL....

GOT TO GO! SEE YOU SOON, DOC!

SEE HOW SIMILAR THE PEUGEOT LOGO IS TO THE STADEL LION-MAN?

PEUGEOT BEGAN AS A SMALL FAMILY BUSINESS IN VALENTIGNEY, A VILLAGE JUST UNDER 200 MILES FROM THE STADEL CAVE!

CLIMB ABOARD, I'LL TAKE YOU THERE.

HERE WE ARE. FRANCE, 1913

AH, GREAT TIMING! THIS IS ARMAND PEUGEOT, THE FOUNDING FATHER, OUT FOR A RIDE...

HE'LL NOTICE MY CAR, FOR SURE.

OH, MY LORD! WHAT THE... WHAT THE... WHAT IS THAT?

WHAT DID I SAY!

GOOD AFTERNOON, MR. PEUGEOT! I'M FROM THE FUTURE, JUST ARRIVED FROM THE YEAR 2020. MY CAR'S A 2020 MODEL OF PEUGEOT!

WELL, I'LL BE....! BUT IT'S HIDEOUS! ARE YOU SURE ABOUT THIS?

OF COURSE. LOOK, THE LION EMBLEM'S STILL THERE.

UPON MY WORD, YOU'RE RIGHT...

BUT... WHEN YOU SAY YOU'RE FROM THE YEAR 2020... ARE YOU TELLING ME MY LITTLE BUSINESS OUTLIVES ME?

OH, AND HOW, MR. PEUGEOT!

IN 2020, THERE ARE PEUGEOT VEHICLES PRETTY MUCH ALL OVER THE WORLD, FROM SYDNEY TO TOKYO! YOUR COMPANY HAS FACTORIES EVERYWHERE—EVEN IN CHINA—AND SELLS MORE THAN 1.5 MILLION CARS EVERY YEAR.

IT EMPLOYS MORE THAN 200,000 PEOPLE, AND HAS AN ANNUAL TURNOVER OF NEARLY 75 BILLION EUROS!

OH LA LA! THIS IS NEWS TO ME! EVEN IN CHINA?

SO, I WAS RIGHT TO LAUNCH INTO AUTOMOBILES! WHEN I INHERITED THE FAMILY BUSINESS, IT WAS A LITTLE WORKSHOP TURNING OUT NOTHING BUT SAWS, SPRINGS AND BICYCLES! I FOUNDED MY AUTOMOBILE COMPANY IN 1896, AND NOW THE WHOLE WORLD HAS HEARD OF PEUGEOT!

YES, YOU STARTED A VERY POWERFUL STORY!

WHAT DO YOU MEAN, A STORY?! THE IMPUDENCE! PEUGEOT ISN'T A STORY!

OKAY, WHAT IS IT, THEN?

WELL, BY ALL THAT'S HOLY, PEUGEOT IS CARS! MILLIONS OF THEM—YOU TOLD ME YOURSELF!

I DID...

BUT JUST IMAGINE... WHAT WOULD HAPPEN TO PEUGEOT IF ALL THE CARS WERE SUDDENLY SENT TO THE SCRAPYARD? BECAUSE OF SOME NEW ENVIRONMENTAL REGULATION, SAY...

AAAH! WHAT A MONSTROUS IDEA!

AHA, BUT THAT WOULDN'T STOP PEUGEOT FACTORIES FROM PRODUCING NEW CARS, WOULD IT, MADAM! WE EVEN HAVE A FACTORY IN CHINA, REMEMBER!

YOU'RE RIGHT, ARMAND. DESTROYING ALL THE VEHICLES WOULDN'T BE THE END OF PEUGEOT...

AND WHAT IF ALL THE WORKERS WENT ON STRIKE AND DEMANDED HIGHER WAGES IN EXCHANGE FOR INCREASED PRODUCTION?

BAH.... THAT'S EASY! PEUGEOT FIRES THEM AND HIRES NEW WORKERS.

ALL SORTED! IN CHINA! I MEAN REALLY!

BUT LET'S IMAGINE SOMETHING EVEN WORSE. WHAT IF A DISASTER STRUCK ALL YOUR FACTORIES WORLDWIDE, DESTROYING ALL THE MACHINERY AND BUILDINGS...?

LORD ABOVE! YOU HAVE THE STRANGEST IDEAS!

WELL, PEUGEOT WOULD TAKE OUT A LOAN, BUY NEW MACHINERY AND BUILD NEW FACTORIES, MADAM!

IT'S DOCTOR ACTUALLY...

OH! I BEG YOUR PARDON.

EXACTLY. SO, PEUGEOT ISN'T THE CARS OR THE WORKERS OR THE FACTORIES...

BUT THEN WHAT IS IT?

IT'S THE MANAGERS AND OWNERS! AND ABOVE ALL, IT'S ME! PEUGEOT IS ME!!!

NO, NO AND NO.

YOU DIE IN 1915, BUT PEUGEOT STILL EXISTS IN 2020. WITH DIFFERENT OWNERS. IN 2014, THE PEUGEOT FAMILY SELLS A LARGE PART OF ITS SHARES TO A CHINESE FIRM. YOU SEE, PEUGEOT DOESN'T JUST HAVE CHINESE FACTORIES, IT ALSO HAS CHINESE OWNERS AND DIRECTORS...

CHINESE OWNERS? WELL, THAT IS A CHANGE! BUT AT LEAST PEUGEOT'S IMMORTAL!

I WOULDN'T GO THAT FAR...

SAY PEUGEOT BREAKS A LAW AND A JUDGE DISSOLVES THE COMPANY. THE FACTORIES WOULD STILL BE THERE, THE WORKERS, ACCOUNTANTS, MANAGERS AND SHAREHOLDERS WOULD STILL BE ALIVE...BUT—BOOM—

PEUGEOT THE COMPANY WOULD BE GONE.

UMMM....

LOOK, I CAN SEE I'VE UPSET YOU. COME ON, ARMAND, I'LL BUY YOU A DRINK IN THE VILLAGE.

WHAT A STRANGE PERSON....

BASICALLY, PEUGEOT THE COMPANY HAS NO ESSENTIAL CONNECTION TO THE PHYSICAL WORLD. SO LET'S COME BACK TO MY FIRST QUESTION: WHAT IS PEUGEOT? DOES YOUR COMPANY REALLY EXIST?

WELL, WELL, WELL... THAT'S A GOOD QUESTION...

AND I CAN ANSWER IT, ARMAND. PEUGEOT IS A FIGMENT OF OUR COLLECTIVE IMAGINATION. WHAT THOSE LAWYERS GUYS CALL A "LEGAL FICTION." IT CAN'T BE POINTED AT, IT'S NOT A PHYSICAL OBJECT!

HUMPH!

YES, BUT IT DOES EXIST AS A LEGAL ENTITY. JUST LIKE YOU AND ME, IT HAS TO OBEY THE LAW, IT CAN OPEN A BANK ACCOUNT, IT CAN OWN PROPERTY...

AND IT PAYS ITS TAXES, DOC!

ABSOLUTELY, AND IT CAN BE SUED, PROSECUTED EVEN, BUT TOTALLY SEPARATELY FROM ANY OF THE PEOPLE WHO OWN OR WORK FOR IT.

PEUGEOT BELONGS TO THE PARTICULAR GENRE OF LEGAL FICTIONS CALLED "LIMITED LIABILITY COMPANIES." THE IDEA BEHIND THESE COMPANIES IS ONE OF HUMANITY'S MOST INGENIOUS INVENTIONS.

HOMO SAPIENS LIVED FOR TENS OF THOUSANDS OF YEARS WITHOUT LIMITED LIABILITY COMPANIES. FOR MOST OF RECORDED HISTORY ONLY FLESH-AND-BLOOD HUMANS COULD OWN PROPERTY!

DON'T I KNOW! WHEN MY ANCESTOR JEAN WAS ALIVE, IN THE 13TH CENTURY, NONE OF THIS WOULD HAVE BEEN POSSIBLE...

VALENTIGNEY, 1213.

TELL ME ABOUT IT! SEE HOW MUCH HARDER THINGS WERE IN HIS DAY... WHEN HE SET UP HIS LITTLE WAGON-MANUFACTURING WORKSHOP, HE WAS THE BUSINESS.

OOOH.... GRANDPA JEAN...

IF A WAGON HE MADE BROKE DOWN A WEEK AFTER HE SOLD IT, THE ANGRY BUYER WOULD SUE JEAN PERSONALLY.

IF HE'D BORROWED 1000 GOLD COINS TO SET UP HIS WORKSHOP AND THE BUSINESS FAILED, HE HAD TO REPAY THE LOAN BY SELLING HIS OWN STUFF—HIS HOUSE, HIS COW, HIS LAND.

HE MIGHT EVEN HAVE HAD TO SELL HIS CHILDREN! IF HE COULDN'T COVER THE DEBT, HE MIGHT HAVE BEEN THROWN IN PRISON. HE WAS 100% LIABLE FOR ALL HIS WORKSHOP'S OBLIGATIONS.

SO IN YOUR GRANDPA JEAN'S DAY, PEOPLE WERE AFRAID TO START NEW BUSINESSES AND TAKE RISKS LIKE THAT. FAILURE COULD MEAN RUIN FOR THEM AND THEIR WHOLE FAMILY.

IF I'D LIVED IN THOSE DAYS, I WOULD PROBABLY HAVE THOUGHT TWICE ABOUT STARTING UP A BUSINESS...

CAFE DU COMMERCE

THE USUAL FOR YOU, MR. ARMAND?

THAT'S WHY PEOPLE INVENTED A REALLY CRAZY STORY—LIMITED LIABILITY COMPANIES! THESE COMPANIES WERE LEGALLY INDEPENDENT OF THE PEOPLE WHO SET THEM UP, OR INVESTED MONEY IN THEM, OR MANAGED THEM.

OVER THE LAST FEW CENTURIES THESE COMPANIES HAVE BECOME THE BIG SHOTS IN THE ECONOMIC ARENA, AND WE'RE SO USED TO THEM, WE FORGET THEY ONLY EXIST IN OUR IMAGINATION.

I DON'T UNDERSTAND ALL OF THIS, BUT I'LL BELIEVE YOU....

IN THE UNITED STATES, THESE LIMITED LIABILITY COMPANIES ARE CALLED "CORPORATIONS"—KIND OF IRONIC, BECAUSE THE WORD COMES FROM "CORPUS," THE LATIN WORD FOR "BODY"... AND THAT'S THE ONE THING THEY DON'T HAVE! SO THEY DON'T HAVE REAL BODIES, BUT THE AMERICAN LEGAL SYSTEM TREATS CORPORATIONS AS LEGAL PERSONS, LIKE THEY WERE FLESH-AND-BLOOD HUMAN BEINGS!

YOU TOOK ADVANTAGE OF THIS SYSTEM YOURSELF, ARMAND. YOU SET UP A COMPANY CALLED PEUGEOT, WHICH IS A DIFFERENT ENTITY FROM YOU.

IF A PEUGEOT BREAKS DOWN, THE BUYER CAN SUE PEUGEOT THE COMPANY, BUT NOT YOU, ARMAND PEUGEOT.

WELL, I SHOULD HOPE NOT!

RIGHT, LET'S GO BACK AND GET OUR CARS.

IF THE COMPANY BORROWED MILLIONS AND THEN WENT BUST, YOU PERSONALLY WOULDN'T OWE A SINGLE FRANC TO THE COMPANY'S CREDITORS.

QUITE RIGHT! IF PEUGEOT THE COMPANY TOOK OUT THE LOAN, THEN IT'S NOTHING TO DO WITH ME!

DO YOU KNOW HOW YOU CREATED THE PEUGEOT COMPANY, ARMAND?

IN MUCH THE SAME WAY THAT PRIESTS AND SORCERERS HAVE CREATED GODS AND DEMONS THROUGHOUT HISTORY...

JESUS, MARY AND JOSEPH! THAT'S BLASPHEMY! I DID NOTHING OF THE SORT!

NO, NO, SERIOUSLY! COMPANIES ARE CREATED IN THE SAME WAY AS GODS—YOU TELL STORIES AND CONVINCE EVERYONE TO BELIEVE THEM.

ACCORDING TO CATHOLIC BELIEFS, IF A CATHOLIC PRIEST WEARING HIS SACRED GARMENTS SOLEMNLY SAYS THE RIGHT WORDS AT THE RIGHT MOMENT, ORDINARY BREAD AND WINE TURN INTO GOD'S FLESH AND BLOOD.

HOC EST CORPUS MEUM!*

*THIS IS MY BODY!

AND, HOCUS POCUS, THE BREAD BECOMES THE FLESH OF GOD. MILLIONS OF DEVOUT CATHOLICS BEHAVE AS IF GOD REALLY IS THERE IN THAT CONSECRATED PIECE OF BREAD.

ACCORDING TO THE FRENCH LEGAL SYSTEM, IF A CERTIFIED LAWYER FOLLOWS ALL THE RIGHT PROCEDURES, WRITING ALL THE RIGHT SPELLS AND OATHS ON A WONDERFULLY DECORATED PIECE OF PAPER, AND PUTTING HIS ORNATE SIGNATURE AT THE BOTTOM OF THE DOCUMENT, THEN HOCUS POCUS—A NEW COMPANY IS INCORPORATED.

ALL DONE!

SO, ARMAND, WHEN YOU WANTED TO SET UP YOUR COMPANY IN 1896, YOU PAID A LAWYER TO GO THROUGH ALL THESE SACRED PROCEDURES. ONCE THE LAWYER HAD PERFORMED HIS RITUALS AND PRONOUNCED HIS MAGIC WORDS, MILLIONS OF FRENCH CITIZENS BEHAVED AS IF THE PEUGEOT COMPANY REALLY EXISTED.

WELL, I NEVER... I'VE NEVER THOUGHT OF IT LIKE THAT...

OKAY, THEN! GOODBYE, DEAR ARMAND!

GOODBYE, DOC!

WELL, I NEVER...

94

TELLING EFFECTIVE STORIES ISN'T EASY.

THE PROBLEM ISN'T TELLING STORIES, IT'S CONVINCING EVERYONE ELSE TO BELIEVE THEM.

MUCH OF HISTORY REVOLVES AROUND ONE BIG QUESTION... HOW DO YOU CONVINCE MILLIONS OF PEOPLE TO BELIEVE A PARTICULAR STORY ABOUT A GOD, A NATION OR A LIMITED LIABILITY COMPANY?

BUT WHEN IT WORKS, IT GIVES SAPIENS SO MUCH POWER BECAUSE IT HELPS MILLIONS OF STRANGERS TO COOPERATE FOR THE SAKE OF COMMON GOALS.

JUST IMAGINE HOW DIFFICULT IT WOULD HAVE BEEN TO ESTABLISH STATES, CHURCHES OR LEGAL SYSTEMS IF WE COULD ONLY TALK ABOUT REAL THINGS LIKE RIVERS, TREES AND LIONS.

OVER THE YEARS, PEOPLE HAVE WOVEN AN INCREDIBLY COMPLEX NETWORK OF STORIES. WITHIN THIS NETWORK, FICTIONS SUCH AS PEUGEOT NOT ONLY EXIST, BUT ALSO ACCUMULATE SERIOUS POWER.

ACADEMICS HAVE NAMES FOR THE THINGS CREATED THROUGH THIS NETWORK OF STORIES. THEY CALL THEM "FICTIONS," "SOCIAL CONSTRUCTS" OR "IMAGINED REALITIES."

AN IMAGINED REALITY ISN'T A LIE.

I'D BE LYING IF I SAID THERE WAS A LION NEAR THE RIVER WHEN I KNEW FOR SURE THERE WASN'T.

THERE'S NOTHING SPECIAL ABOUT LIES. GREEN MONKEYS AND CHIMPANZEES CAN LIE TOO. A GREEN MONKEY'S BEEN SEEN DOING THIS...

KYAK! KYAK! KYAK!*

*WATCH OUT, A LION!

UNLIKE A LIE, AN IMAGINED REALITY IS SOMETHING THAT EVERYONE BELIEVES IN, AND AS LONG AS THEY CONTINUE TO BELIEVE IN IT, THE IMAGINED REALITY CAN EXERT A LOT OF POWER.

PERHAPS THE SCULPTOR FROM STADEL CAVE SINCERELY BELIEVED IN A LION-MAN GUARDIAN SPIRIT. SOME SORCERERS ARE CHARLATANS, BUT MOST SINCERELY BELIEVE IN THE EXISTENCE OF GODS AND DEMONS.

MOST MILLIONAIRES SINCERELY BELIEVE IN THE EXISTENCE OF MONEY AND LIMITED LIABILITY COMPANIES. MOST HUMAN RIGHTS ACTIVISTS SINCERELY BELIEVE IN THE EXISTENCE OF HUMAN RIGHTS.

NOBODY LIED WHEN, IN 2011, THE UNITED NATIONS DEMANDED THAT THE LIBYAN GOVERNMENT RESPECT ITS CITIZENS' HUMAN RIGHTS, EVEN THOUGH THE UNITED NATIONS, LIBYA AND HUMAN RIGHTS ARE ALL FIGMENTS OF OUR CRAZILY FERTILE IMAGINATION.

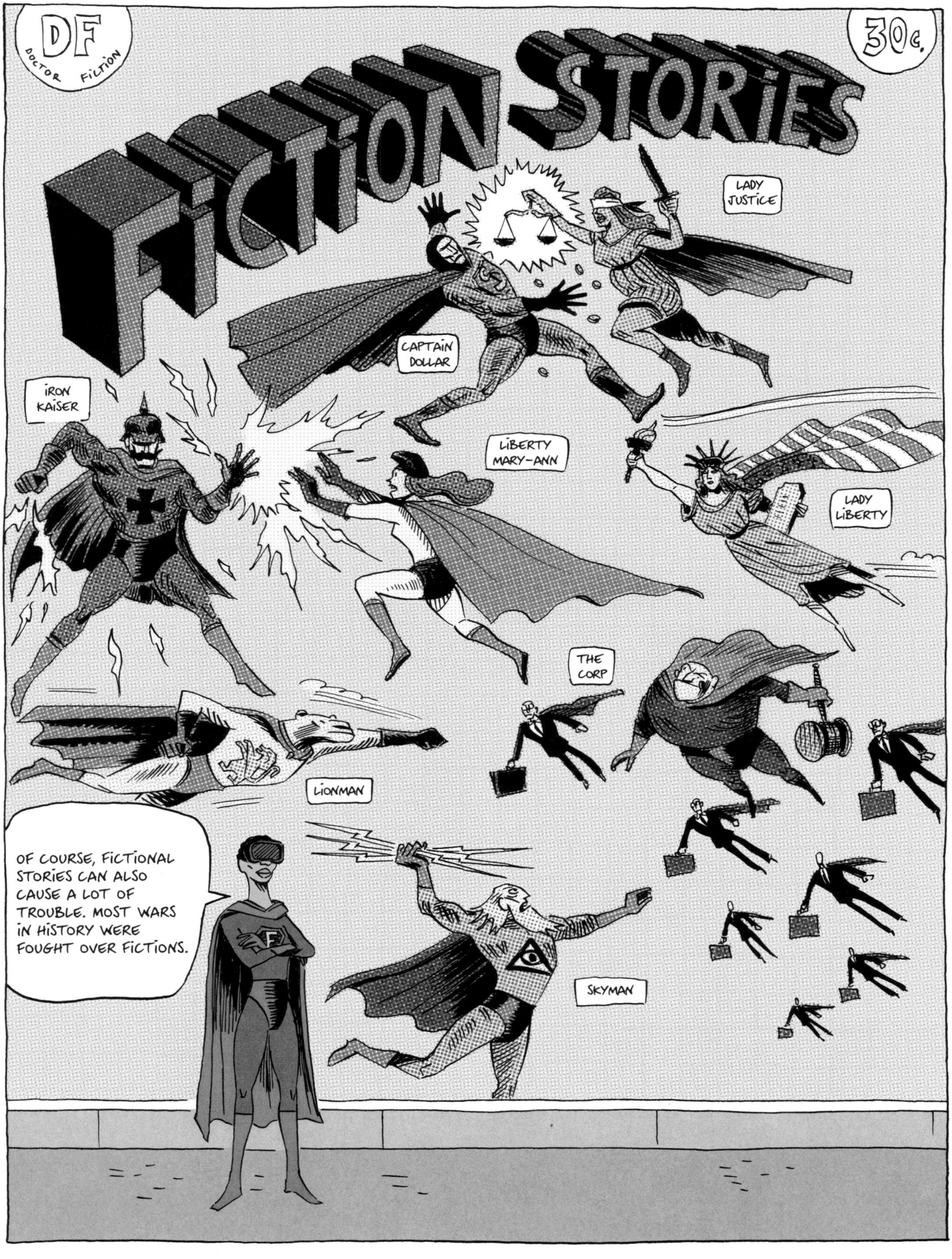
DF
DOCTOR FICTION
30¢
FICTION STORIES
LADY JUSTICE
CAPTAIN DOLLAR
IRON KAISER
LIBERTY MARY-ANN
LADY LIBERTY
THE CORP
LIONMAN
OF COURSE, FICTIONAL STORIES CAN ALSO CAUSE A LOT OF TROUBLE. MOST WARS IN HISTORY WERE FOUGHT OVER FICTIONS.
SKYMAN

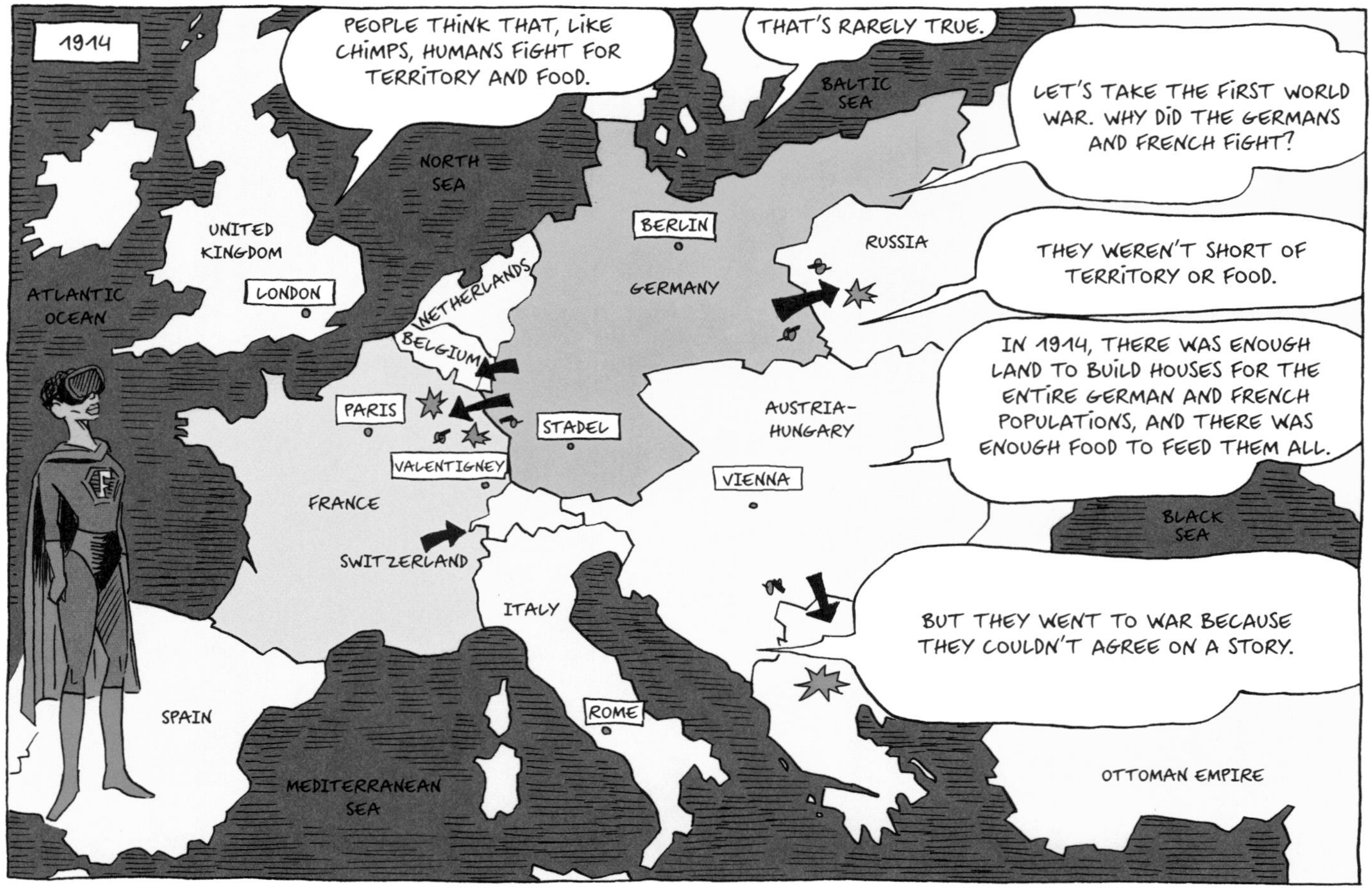
1914
PEOPLE THINK THAT, LIKE CHIMPS, HUMANS FIGHT FOR TERRITORY AND FOOD.
THAT'S RARELY TRUE.
LET'S TAKE THE FIRST WORLD WAR. WHY DID THE GERMANS AND FRENCH FIGHT?
THEY WEREN'T SHORT OF TERRITORY OR FOOD.
IN 1914, THERE WAS ENOUGH LAND TO BUILD HOUSES FOR THE ENTIRE GERMAN AND FRENCH POPULATIONS, AND THERE WAS ENOUGH FOOD TO FEED THEM ALL.
BUT THEY WENT TO WAR BECAUSE THEY COULDN'T AGREE ON A STORY.
BALTIC SEA
NORTH SEA
UNITED KINGDOM
LONDON
ATLANTIC OCEAN
NETHERLANDS
BELGIUM
BERLIN
GERMANY
RUSSIA
PARIS
STADEL
VALENTIGNEY
AUSTRIA-HUNGARY
VIENNA
FRANCE
BLACK SEA
SWITZERLAND
ITALY
SPAIN
ROME
MEDITERRANEAN SEA
OTTOMAN EMPIRE

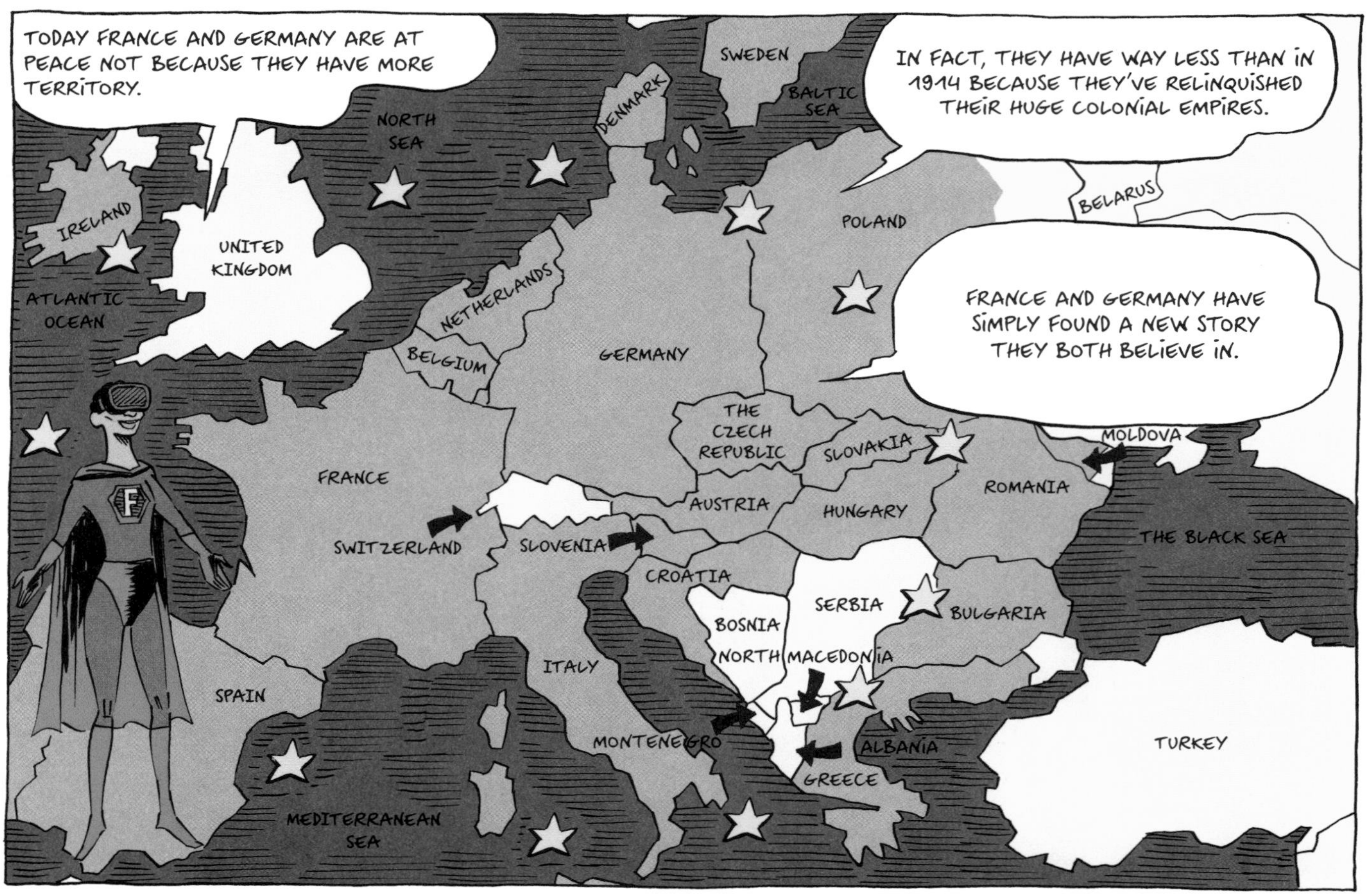
TODAY FRANCE AND GERMANY ARE AT PEACE NOT BECAUSE THEY HAVE MORE TERRITORY.
IN FACT, THEY HAVE WAY LESS THAN IN 1914 BECAUSE THEY'VE RELINQUISHED THEIR HUGE COLONIAL EMPIRES.
FRANCE AND GERMANY HAVE SIMPLY FOUND A NEW STORY THEY BOTH BELIEVE IN.
SWEDEN
DENMARK
BALTIC SEA
NORTH SEA
BELARUS
IRELAND
UNITED KINGDOM
POLAND
ATLANTIC OCEAN
NETHERLANDS
BELGIUM
GERMANY
THE CZECH REPUBLIC
SLOVAKIA
MOLDOVA
FRANCE
AUSTRIA
ROMANIA
HUNGARY
THE BLACK SEA
SWITZERLAND
SLOVENIA
CROATIA
SERBIA
BOSNIA
BULGARIA
NORTH MACEDONIA
ITALY
SPAIN
MONTENEGRO
ALBANIA
TURKEY
GREECE
MEDITERRANEAN SEA

HUMANS ARE REALLY WEIRD ANIMALS.

SO, BASICALLY EVERYTHING'S FICTION?

NO, NO... REALITY'S STILL THERE.

WHEN SOMEONE'S KILLED IN A WAR, IT'S VERY REAL!

FICTIONS ARE SUPER IMPORTANT BECAUSE THEY HELP US COOPERATE. IF PEOPLE DIDN'T BELIEVE IN MONEY AND CORPORATIONS, THE GLOBAL TRADE NETWORK WOULD COLLAPSE.

IF PEOPLE DIDN'T BELIEVE IN NATIONS, THEY WOULDN'T PAY TAXES...

...THEN THERE'D BE NO NATIONAL SYSTEMS OF EDUCATION AND HEALTHCARE.

BUT SAPIENS SHOULD NEVER FORGET THAT FICTIONS ARE JUST TOOLS! WE DREAMED THEM UP TO SERVE OUR NEEDS!

PEOPLE MUSTN'T BECOME SLAVES TO THEIR OWN TOOLS! IF PEOPLE FORGET THIS, AND START WAGING WARS FOR THE PROFITS OF A COMPANY OR FOR THE GLORY OF A NATION, THAT WOULD BE TRULY TERRIBLE!

BUT HOW CAN WE TELL IF THE HERO IN A STORY IS REAL OR INVENTED?

JUST ASK YOURSELF: CAN THE HERO SUFFER?

A CORPORATION CAN'T SUFFER, EVEN IF IT GOES BANKRUPT. IT DOESN'T HAVE A MIND, AND CAN'T FEEL PAIN OR SADNESS.

HO HO HO! DOESN'T EVEN HURT!

AND A NATION CAN'T SUFFER, EVEN IF IT LOSES A WAR.

BUT A HUMAN BEING WOUNDED IN CONFLICT REALLY SUFFERS.

WHICH IS WHY WE SHOULD BE VERY CAREFUL NOT TO MAKE REAL PEOPLE SUFFER FOR THE SAKE OF A FICTION.

YES, OF COURSE.

BUT THERE'S SOMETHING I STILL DON'T GET...

WHAT'S THAT, ZOE?

HOW COME SAPIENS HAVE THIS ABILITY TO INVENT FICTION?

THE MOST LIKELY EXPLANATION IS THAT ACCIDENTAL GENETIC MUTATIONS CHANGED THE WIRING IN OUR BRAINS...

BLA BLA BLA BLA

...WHICH MEANT SAPIENS COULD THINK IN COMPLETELY NEW WAYS AND COMMUNICATE USING A WHOLE NEW TYPE OF LANGUAGE.

AND WHY DID THIS DNA MUTATION HAPPEN TO SAPIENS AND NOT NEANDERTHALS?

PURE CHANCE, AS FAR AS WE KNOW.

BUT IT'S A PITY WE DON'T KNOW FOR SURE, RIGHT?

IF YOU LIKE, I COULD GIVE YOU ALL KINDS OF WAY MORE INTERESTING EXPLANATIONS.

IT WAS A GENETIC EXPERIMENT CARRIED OUT BY ALIENS FROM ALPHA CENTAURI. YOU KNOW, THE ALIENS WHO...

THANK YOU, THANK YOU, DOCTOR FICTION, BUT SCIENCE IS PROBABLY THE ONE AREA WHERE WE'RE BETTER OFF WITHOUT YOUR HELP.

IN SCIENCE, IT'S BETTER TO ADMIT YOU DON'T KNOW THAN TO INVENT FICTIONS.

OKAY, OKAY... I WAS ONLY TRYING TO BE HELPFUL...

RIGHT! MY FRIENDS AND I STILL HAVE LOTS OF WORK TO DO. WE'RE OUT OF HERE!

KEEP UP THE GOOD WORK, MAYBE SEE YOU AGAIN SOON!

SURE THING!

BYE!

THAT'S WHY WE SAPIENS ARE SO GOOD AT COOPERATING:

WE HAVE UNIQUE COMMUNICATION ABILITIES. WE CAN TALK ABOUT LIONS AND HUMANS, AND WE CAN EVEN INVENT STORIES ABOUT FICTIONAL LION-MEN.

ALL THIS RELIES ON VERY SPECIAL COGNITIVE SKILLS— INVENTING, REMEMBERING, LEARNING AND COMMUNICATING.

THESE COGNITIVE SKILLS APPEARED ABOUT 70,000 YEARS AGO, IN WHAT'S KNOWN AS THE COGNITIVE REVOLUTION

EVER SINCE THEN, SAPIENS HAVE BEEN LIVING IN A DUAL REALITY.

ON THE ONE HAND, THE OBJECTIVE REALITY OF RIVERS, TREES AND LIONS...

SAFARI
TANZANIA

ON THE OTHER, THE IMAGINED REALITY OF GODS, NATIONS AND COMPANIES.

All other passports

UK/EU passports

AS TIME WENT BY, THE IMAGINED REALITY BECAME MORE AND MORE POWERFUL...

SO THAT NOW THE VERY SURVIVAL OF RIVERS, TREES AND LIONS DEPENDS ON THE GOOD GRACE OF IMAGINARY ENTITIES SUCH AS ALMIGHTY GODS, THE UNITED STATES OR GOOGLE...

WAKE UP! TIME IS RUNNING OUT!

HAH! MORE FAKE NEWS!

MAKE AMER GREAT AGAI

THANKS FOR THE TRIP, UNCLE YUVAL!

BYE, ZOE!

OUR ABILITY TO CREATE AN IMAGINED REALITY OUT OF WORDS MADE IT POSSIBLE FOR LARGE NUMBERS OF STRANGERS TO COOPERATE EFFECTIVELY.

BUT IT DID MORE THAN THAT.

SINCE LARGE-SCALE HUMAN COOPERATION IS BASED ON MYTHS, THE WAY PEOPLE COOPERATE CAN BE ALTERED VERY QUICKLY BY CHANGING THE MYTHS—BY TELLING DIFFERENT STORIES.

VERSAILLES, 1788.

PARIS, 1793.

DURING THE FRENCH REVOLUTION, MILLIONS OF PEOPLE CHANGED THEIR BELIEFS ALMOST OVERNIGHT. THEY ABANDONED THE MYTH OF THE DIVINE RIGHT OF KINGS...

FOR THE MYTH OF THE SOVEREIGNTY OF THE PEOPLE.

SO EVER SINCE THE COGNITIVE REVOLUTION HOMO SAPIENS HAS BEEN ABLE TO REVISE ITS BEHAVIOR RAPIDLY, ADAPTING TO CHANGING NEEDS. THIS OPENED A FAST LANE OF CULTURAL EVOLUTION, BYPASSING THE TRAFFIC JAMS OF GENETIC EVOLUTION.

ONLY WE SAPIENS CAN CHANGE OUR SOCIAL SYSTEM SO QUICKLY, WOULDN'T YOU SAY, PROFESSOR SARASWATI?

ABSOLUTELY, YUVAL.

HEY GUYS, SEE YOU LATER! HA! HA! HA!

BEHAVIOR IN OTHER SOCIAL ANIMALS IS LARGELY DETERMINED BY THEIR GENES.

OF COURSE, DNA ISN'T AN AUTOCRAT. ANIMAL BEHAVIOR IS ALSO INFLUENCED BY ENVIRONMENTAL FACTORS AND INDIVIDUAL QUIRKS.

STILL, IN A GIVEN ENVIRONMENT, ANIMALS OF THE SAME SPECIES WILL TEND TO BEHAVE IN A SIMILAR WAY.

AND SIGNIFICANT CHANGES IN SOCIAL BEHAVIOR DON'T GENERALLY HAPPEN WITHOUT GENETIC MUTATIONS.

PYGMY CHIMPANZEES—ALSO KNOWN AS BONOBOS—USUALLY LIVE IN EGALITARIAN GROUPS DOMINATED BY FEMALE ALLIANCES.

WHILE COMMON CHIMPANZEES LIVE IN HIERARCHICAL GROUPS HEADED BY AN ALPHA MALE.

FEMALE COMMON CHIMPANZEES CAN'T LEARN FROM THEIR BONOBO RELATIVES AND STAGE A FEMINIST REVOLUTION!

DIVERSITY. EQUALITY. UNITY.

MALE CHIMPS CAN'T GATHER IN A CONSTITUTIONAL ASSEMBLY, ABOLISH THE ROLE OF ALPHA MALE AND DECLARE THAT AS OF NOW ALL CHIMPS ARE TO BE TREATED AS EQUALS.

DRAMATIC CHANGES IN BEHAVIOR LIKE THAT WOULD ONLY HAPPEN IF SOMETHING CHANGED IN THE CHIMPANZEES' DNA.

THANKS FOR THE EXPLANATION, PROFESSOR SARASWATI.

FOR SIMILAR REASONS, ARCHAIC HUMANS DIDN'T INITIATE ANY REVOLUTIONS.

AS FAR AS WE CAN TELL, CHANGES IN SOCIAL PATTERNS, THE INVENTION OF NEW TECHNOLOGIES, AND THE SETTLEMENT OF ALIEN HABITATS RESULTED FROM GENETIC MUTATIONS AND ENVIRONMENTAL PRESSURES MORE THAN FROM CULTURAL INITIATIVES.

I HEREBY DECLARE THAT SURFING IS HOT THIS SUMMER!

LET'S GO, DUDES!

?

THAT'S WHY IT TOOK ARCHAIC HUMANS HUNDREDS OF THOUSANDS OF YEARS TO MAKE THESE CHANGES.

TWO MILLION YEARS AGO, GENETIC MUTATIONS PRODUCED A NEW HUMAN SPECIES CALLED HOMO ERECTUS. A NEW STONE TOOL TECHNOLOGY EMERGED AT THE SAME TIME, AND IS NOW RECOGNIZED AS A DEFINING FEATURE OF THIS SPECIES.

OKAY, OKAY... IT WAS JUST A SUGGESTION...

AS LONG AS HOMO ERECTUS DIDN'T UNDERGO FURTHER GENETIC ALTERATIONS, ITS STONE TOOLS REMAINED ROUGHLY THE SAME... FOR CLOSE TO TWO MILLION YEARS!

-2 MILLION YEARS

TODAY

HOMO ERECTUS

0 1 2 3 4 5 6 7 8 9 10 11 12 13 14 15 16 17 18 19 20

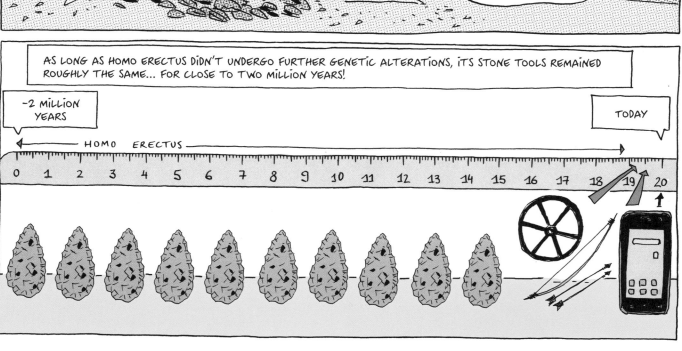

IN CONTRAST, EVER SINCE THE COGNITIVE REVOLUTION, SAPIENS HAVE BEEN ABLE TO CHANGE THEIR BEHAVIOR QUICKLY, TRANSMITTING NEW BEHAVIORS TO FUTURE GENERATIONS WITHOUT NEEDING GENETIC OR ENVIRONMENTAL CHANGES.

HERE'S A PRIME EXAMPLE: HUMANS SOMETIMES HAVE CHILDLESS ELITES.

EUNUCH OF THE FORBIDDEN CITY

BUDDHIST NUN

CATHOLIC PRIEST

THE VERY EXISTENCE OF THESE ELITES GOES AGAINST THE MOST FUNDAMENTAL PRINCIPLES OF NATURAL SELECTION, BECAUSE THESE DOMINANT MEMBERS OF SOCIETY WILLINGLY GIVE UP PROCREATION.

WHILE A CHIMPANZEE ALPHA MALE USES HIS POWER TO HAVE SEX WITH AS MANY FEMALES AS POSSIBLE— THEREFORE SIRING A LARGE PROPORTION OF HIS TROOP'S YOUNG—THE CATHOLIC ALPHA MALE ABSTAINS COMPLETELY FROM SEXUAL INTERCOURSE OR RAISING A FAMILY.

GRUNT! GRUNT! GRUNT!

??

PLANETV

THIS ABSTINENCE DOESN'T RESULT FROM UNIQUE ENVIRONMENTAL CONDITIONS SUCH AS A SEVERE LACK OF FOOD OR OF POTENTIAL MATES.

NOR IS IT DUE TO SOME QUIRKY GENETIC MUTATION. THE CATHOLIC CHURCH HAS SURVIVED FOR CENTURIES, NOT BY PASSING ON A "CELIBACY GENE" FROM ONE POPE TO THE NEXT...

BUT BY PASSING ON THE STORIES OF THE BIBLE AND OF CATHOLIC CANON LAW.

THE BIBLE

IN OTHER WORDS, WHILE THE BEHAVIOR PATTERNS OF ARCHAIC HUMANS STAYED THE SAME FOR TENS OF THOUSANDS OF YEARS, SAPIENS COULD TRANSFORM THEIR SOCIAL STRUCTURES, THEIR INTERPERSONAL RELATIONS, THEIR ECONOMIC ACTIVITIES AND A HOST OF OTHER BEHAVIORS WITHIN A COUPLE OF DECADES!

1914:
SECOND REICH.

LET'S LOOK AT THE LIFE OF A WOMAN BORN IN BERLIN IN 1900...

1924:
THE WEIMAR REPUBLIC.

1936: THIRD REICH.

1961: THE BERLIN WALL SEPARATES EAST AND WEST GERMANY.

DANGER!
BORDER!

1989: GERMAN REUNIFICATION.

SHE MANAGED TO EXPERIENCE FIVE DIFFERENT REGIMES, EVEN THOUGH HER DNA REMAINED EXACTLY THE SAME.

SAPIENS

NEANDERTHAL

THE COGNITIVE REVOLUTION WAS THE KEY TO SAPIENS' SUCCESS.

IN A ONE-ON-ONE BRAWL, A NEANDERTHAL WOULD PROBABLY BEAT A SAPIENS.

BUT IN A LARGE-SCALE CONFLICT, NEANDERTHALS DIDN'T STAND A CHANCE.

WHY DO WE KEEP LOSING? I MEAN, WE'RE BIGGER AND STRONGER!

AND WE CAN TELL EACH OTHER WHAT THE LIONS ARE DOING TOO, YOU KNOW!

YES, BUT NOT AS WELL AS THE SAPIENS... AND YOU PROBABLY CAN'T TELL STORIES ABOUT TRIBAL GUARDIAN SPIRITS.

HUH? ABOUT WHAT?

STORIES ABOUT TRIBAL SPIRITS CAN HELP YOU... TO TRUST STRANGERS, FOR EXAMPLE.

TRUST STRANGERS?! WHOAH, THAT'S DANGEROUS!

BUT IF YOU DON'T TRUST STRANGERS, HOW DO YOU TRADE WITH THEM?

TRADE? WHAT'S TRADE?

IT'S REALLY IMPORTANT.

AND COULD YOUR SAPIENS "TRADE" 30,000 YEARS AGO?

YES. ARCHAEOLOGISTS EXCAVATING 30,000-YEAR-OLD SAPIENS SITES IN CENTRAL EUROPE OCCASIONALLY FIND SEASHELLS THAT WERE BROUGHT HUNDREDS OF MILES FROM THE MEDITERRANEAN AND ATLANTIC COASTS.

WELL, WE MAKE OUR TOOLS WITH LOCAL MATERIALS!

SEASHELLS ARE USELESS, ANYWAY!

AH, BUT THOSE SEASHELLS ARE EVIDENCE THAT DIFFERENT SAPIENS BANDS TRADED WITH ONE ANOTHER.

THERE'S NO EVIDENCE OF TRADE LIKE THIS ON YOUR SITES.

PAH! SEASHELLS ARE FOR KIDS!

OK, SO HERE'S ANOTHER EXAMPLE.

LOOK AT THIS MAP OF THE SOUTH PACIFIC. BANDS OF SAPIENS LIVED ON THE SMALL ISLAND OF NEW IRELAND, ABOUT 30,000 YEARS AGO.

PAPUA-NEW-GUINEA

NEW IRELAND

AUSTRALIA

THESE SAPIENS USED A VOLCANIC GLASS CALLED OBSIDIAN TO MAKE SERIOUSLY STRONG, SHARP TOOLS.

OUCH! THAT IS SHARP!

BUT NEW IRELAND HAS NO NATURAL DEPOSITS OF OBSIDIAN. LAB TESTS SHOWED THAT THE OBSIDIAN THEY USED WAS BROUGHT FROM DEPOSITS 250 MILES AWAY.

NEW IRELAND

NEW BRITAIN

HERE, ON THE ISLAND OF NEW BRITAIN.

SOME SAPIENS ON THESE ISLANDS MUST HAVE BEEN EXCELLENT SAILORS WHO TRADED FROM ISLAND TO ISLAND OVER LONG DISTANCES.

SAILORS? WHAT'S THE POINT OF THAT?

WELL, TO TRADE, FOR A START.

TRADE, TRADE... IS THAT ALL YOU CAN THINK ABOUT?

I KNOW TRADE MAY SEEM VERY PRAGMATIC. IT SHOULDN'T NEED FICTIONAL STORIES, BUT ALL TRADE NETWORKS IN HISTORY HAVE BEEN BASED ON FICTIONS.

I JUST DON'T GET THIS STUFF! COME ON, GRAOW, LET'S SPLIT, HE'S MESSING WITH OUR HEADS!

YOU'RE RIGHT, AND I DON'T TRUST STRANGERS, ANYWAY.

NO, PLEASE DON'T GO!

THAT'S EXACTLY THE PROBLEM, MY FRIENDS! TRADE CAN'T EXIST WITHOUT TRUST, WHICH IS WHY SAPIENS IS THE ONLY ANIMAL EVER TO USE IT. TRUSTING STRANGERS ISN'T EASY.

YOU'RE NOT KIDDING!

KEEP TALKING!

TODAY'S GLOBAL TRADE NETWORK IS BASED ON OUR TRUST IN FICTIONAL ENTITIES LIKE THE TOTEMS OF FINANCE.

 WHEN TWO STRANGERS IN A TRIBAL SOCIETY WANT TO TRADE, THEY ESTABLISH TRUST BY APPEALING TO A COMMON GOD, A MYTHICAL ANCESTOR OR A TOTEM ANIMAL.

IT'S EXACTLY THE SAME TODAY. TRADE BETWEEN STRANGERS IN THE TWENTY-FIRST CENTURY ALSO DEPENDS ON APPEALING TO COMMON GODS, ANCESTORS AND TOTEMS.

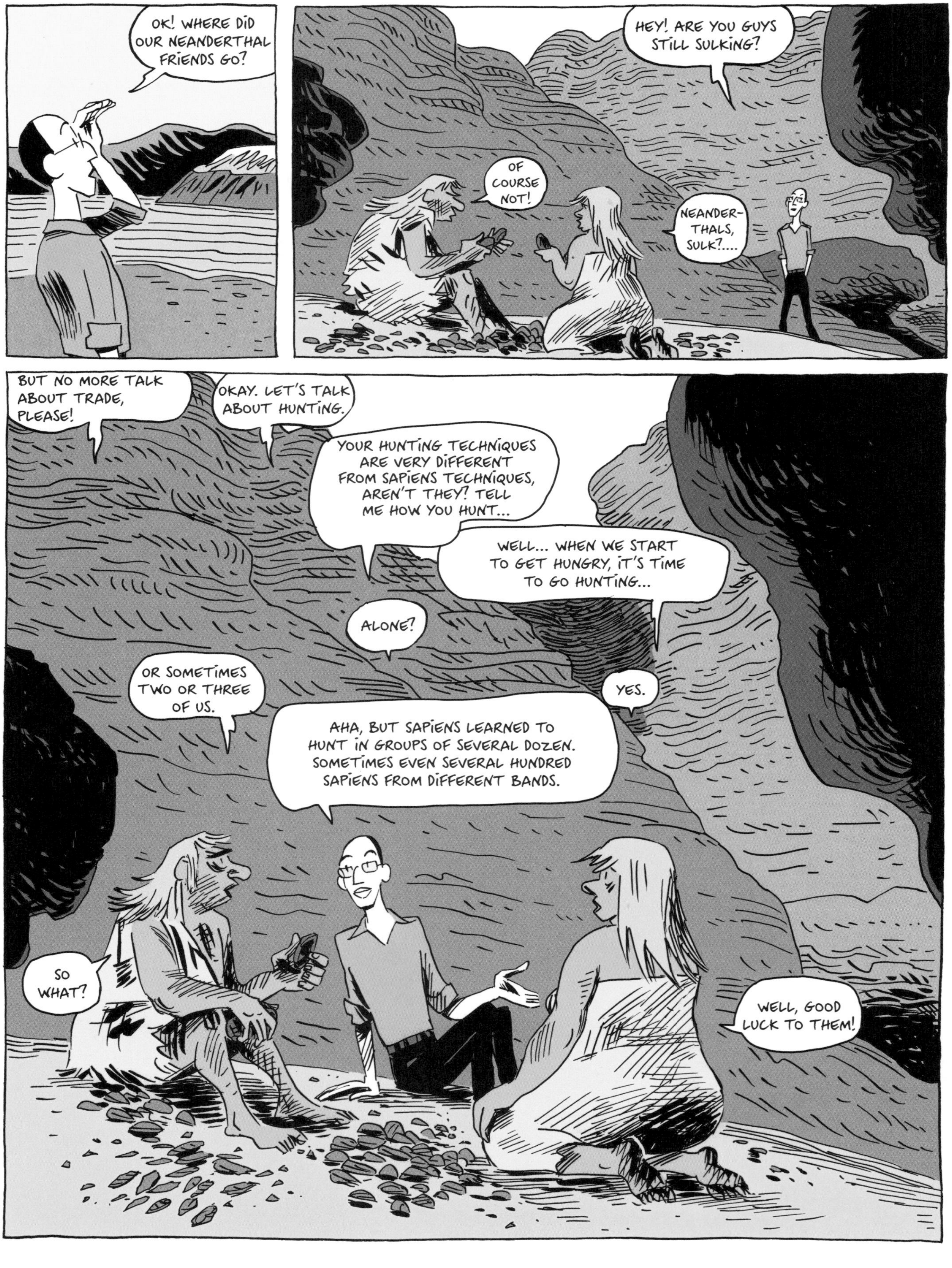
OK! WHERE DID OUR NEANDERTHAL FRIENDS GO?
HEY! ARE YOU GUYS STILL SULKING?
OF COURSE NOT!
NEANDER-THALS, SULK?....
BUT NO MORE TALK ABOUT TRADE, PLEASE!
OKAY. LET'S TALK ABOUT HUNTING.
YOUR HUNTING TECHNIQUES ARE VERY DIFFERENT FROM SAPIENS TECHNIQUES, AREN'T THEY? TELL ME HOW YOU HUNT...
WELL... WHEN WE START TO GET HUNGRY, IT'S TIME TO GO HUNTING...
ALONE?
YES.
OR SOMETIMES TWO OR THREE OF US.
AHA, BUT SAPIENS LEARNED TO HUNT IN GROUPS OF SEVERAL DOZEN. SOMETIMES EVEN SEVERAL HUNDRED SAPIENS FROM DIFFERENT BANDS.
SO WHAT?
WELL, GOOD LUCK TO THEM!

IN LARGER NUMBERS, THEY COULD SURROUND AN ENTIRE HERD OF ANIMALS.

THEY WORKED TOGETHER, TRAPPING THE HERD BY A RIVER OR IN A NARROW GORGE.

THEY COULD KILL AN ENTIRE HERD IN ONE AFTERNOON.

YEAH, AND DON'T WE KNOW IT! THEY WERE SO PUSHY THEY EVEN DID IT ON OUR TERRITORY!

SO THEN WE DIDN'T HAVE ENOUGH TO EAT!

SAPIENS SOMETIMES EVEN BUILT FENCES AND DUG HOLES TO TRAP ANIMALS.

UGH... THAT'S A LOT OF WORK! NO BAND COULD DO THAT ON ITS OWN...

SO WHY NOT ASK NEIGHBORING BANDS TO JOIN YOU?

NEIGHBORING BANDS! BUT THEY'RE STRANGERS! WE CAN'T TRUST THEM—AND THEY DON'T TRUST US.

WELL, YOU COULD TELL THEM YOU HAD A VISION OF THE GREAT LION-SPIRIT IN THE SKY ORDERING ALL NEANDERTHALS TO JOIN FORCES AGAINST THE SAPIENS.

A GREAT LION SPIRIT! IN THE SKY! WHATEVER!

ANYWAY, A LION CAN'T FLY. IT WOULD FALL DOWN!

DID YOU EVER HEAR ANYTHING SO DUMB?

NOT REALLY!

HA! HA!

HA! HA!

BUT, WHAT SHOULD WE DO ABOUT THESE SAPIENS?

UNTIL THE COGNITIVE REVOLUTION, SAPIENS WERE JUST ANOTHER SPECIES OF ANIMAL. THEY BELONGED TO THE REALM OF BIOLOGY, AND WE USE BIOLOGICAL THEORIES TO UNDERSTAND THEM.

AFTER THE COGNITIVE REVOLUTION, SAPIENS INVENTED LOTS OF FICTIONAL STORIES... AND STARTED LOTS OF STRANGE NEW BEHAVIOR PATTERNS, ESTABLISHING WHAT WE CALL "CULTURES."

HISTORY BEGINS WITH THE COGNITIVE REVOLUTION. OUR PRIMARY MEANS OF EXPLAINING THE DEVELOPMENT OF HOMO SAPIENS FROM THEN ON ARE HISTORICAL NARRATIVES RATHER THAN BIOLOGICAL THEORIES.

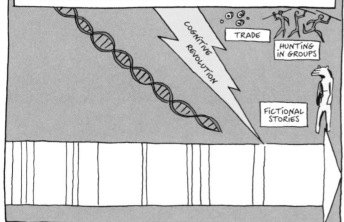

THIS DOESN'T MEAN HOMO SAPIENS AND HUMAN CULTURE WERE SUDDENLY EXEMPT FROM BIOLOGICAL LAWS. WE'RE STILL ANIMALS, AND OUR PHYSICAL, EMOTIONAL AND COGNITIVE ABILITIES ARE STILL SHAPED BY OUR DNA.

PAUL'S CONVERSION ON THE ROAD TO DAMASCUS

BUT TO UNDERSTAND THE RISE OF CHRISTIANITY OR THE FRENCH REVOLUTION, IT'S NOT ENOUGH TO KNOW HOW GENES, HORMONES AND ORGANISMS INTERACT. YOU HAVE TO TAKE INTO ACCOUNT HOW IDEAS, IMAGES AND FANTASIES INTERACT AS WELL.

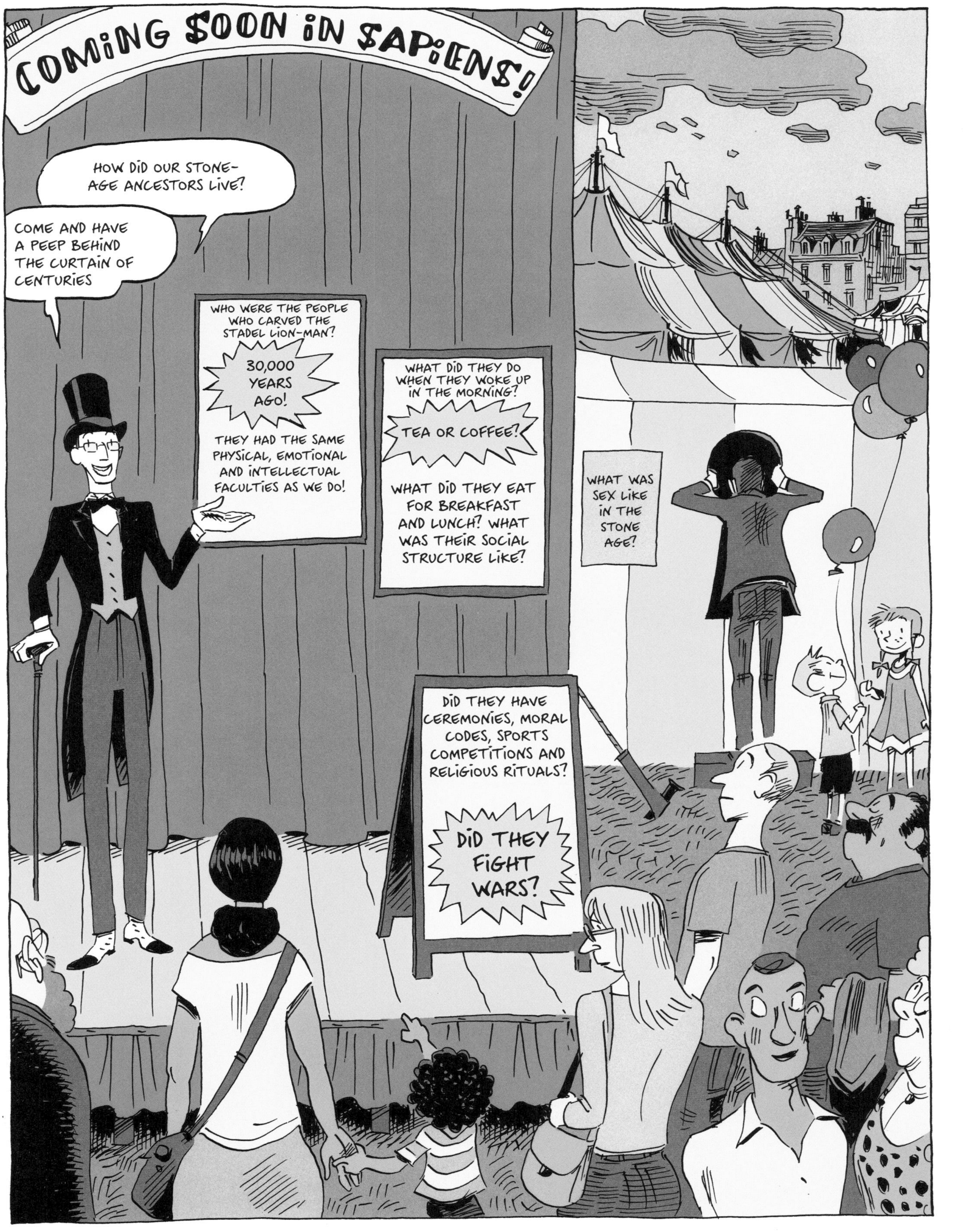
COMING SOON IN SAPIENS!
HOW DID OUR STONE-AGE ANCESTORS LIVE?
COME AND HAVE A PEEP BEHIND THE CURTAIN OF CENTURIES
WHO WERE THE PEOPLE WHO CARVED THE STADEL LION-MAN?
30,000 YEARS AGO!
THEY HAD THE SAME PHYSICAL, EMOTIONAL AND INTELLECTUAL FACULTIES AS WE DO!
WHAT DID THEY DO WHEN THEY WOKE UP IN THE MORNING?
TEA OR COFFEE?
WHAT DID THEY EAT FOR BREAKFAST AND LUNCH? WHAT WAS THEIR SOCIAL STRUCTURE LIKE?
WHAT WAS SEX LIKE IN THE STONE AGE?
DID THEY HAVE CEREMONIES, MORAL CODES, SPORTS COMPETITIONS AND RELIGIOUS RITUALS?
DID THEY FIGHT WARS?

SEX, LIES AND
CAVE PAINTERS

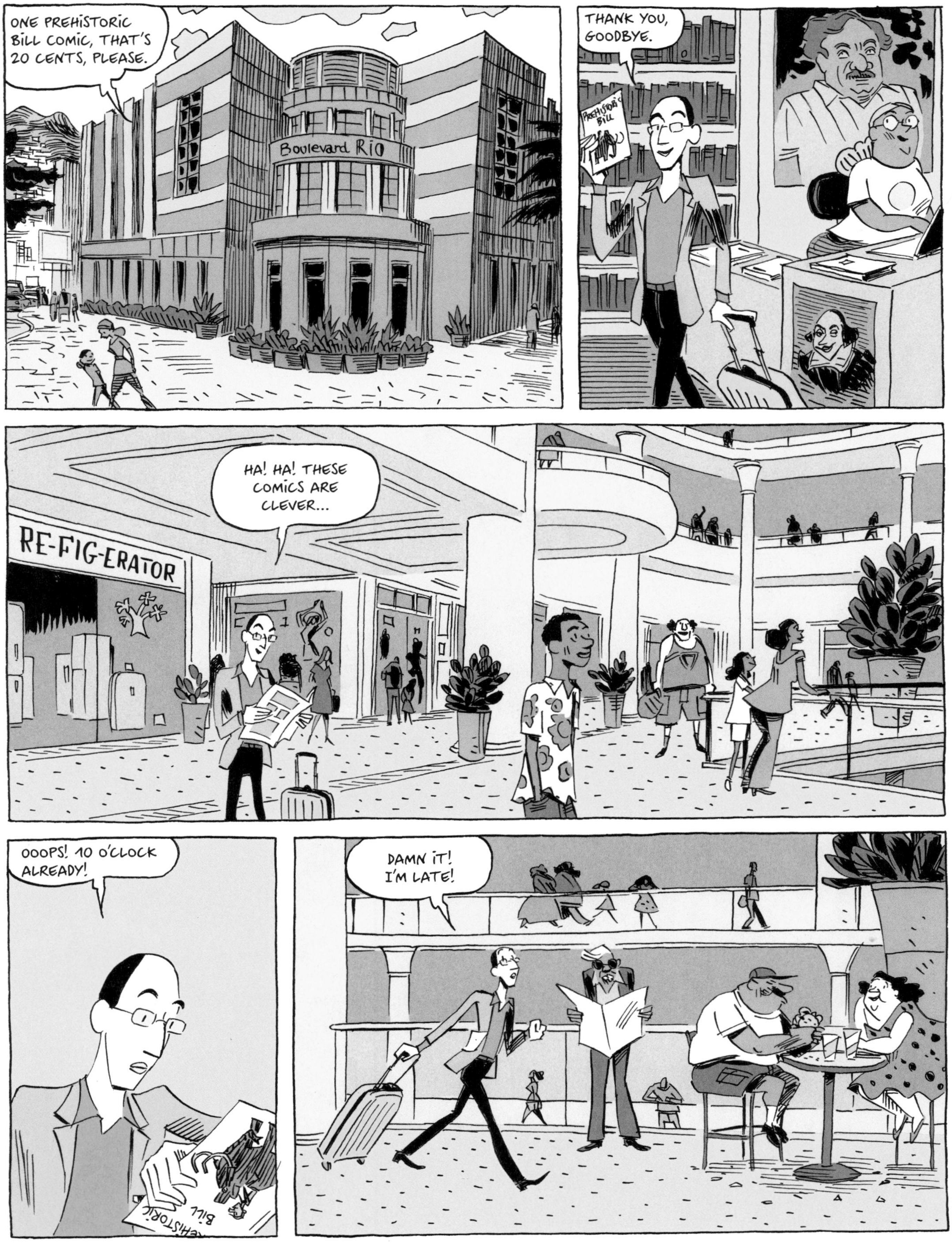
ONE PREHISTORIC BILL COMIC, THAT'S 20 CENTS, PLEASE.
Boulevard RIO
THANK YOU, GOODBYE.
PREHISTORIC BILL
HA! HA! THESE COMICS ARE CLEVER...
RE-FIG-ERATOR
OOOPS! 10 O'CLOCK ALREADY!
DAMN IT! I'M LATE!

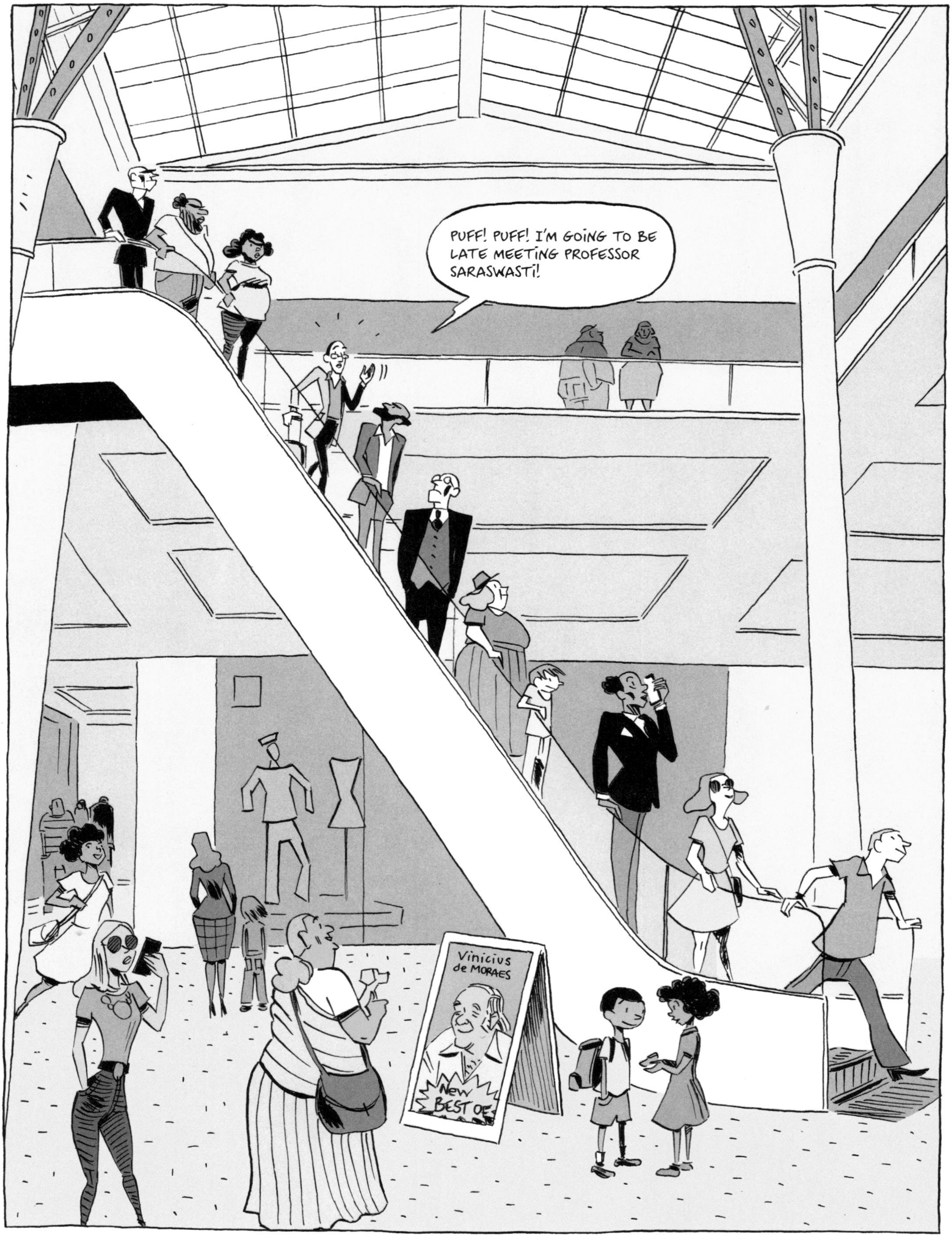
PUFF! PUFF! I'M GOING TO BE LATE MEETING PROFESSOR SARASWASTI!
Vinicius de MORAES
New BEST OF

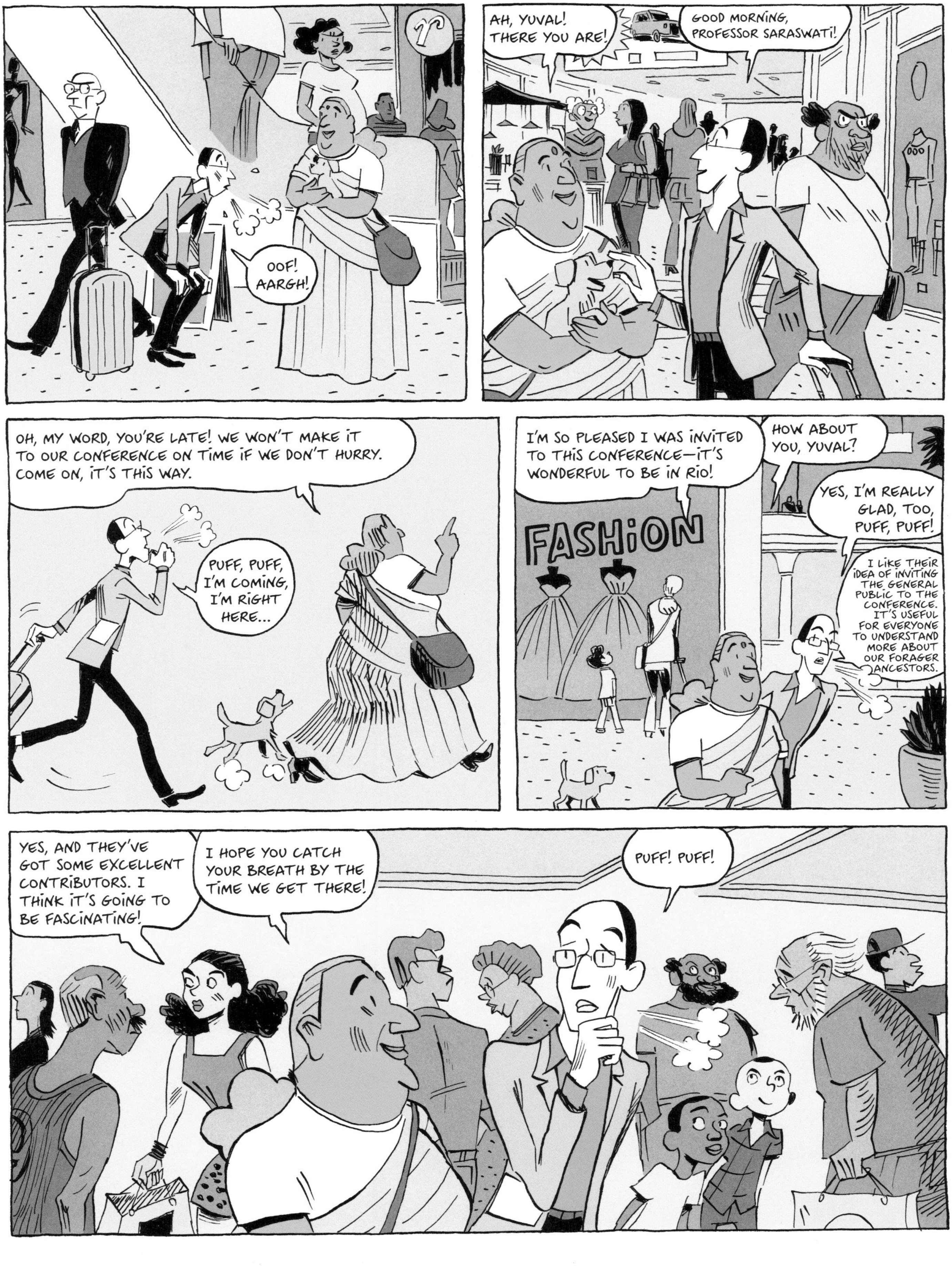
OOF! AARGH!
AH, YUVAL! THERE YOU ARE!
GOOD MORNING, PROFESSOR SARASWATI!
OH, MY WORD, YOU'RE LATE! WE WON'T MAKE IT TO OUR CONFERENCE ON TIME IF WE DON'T HURRY. COME ON, IT'S THIS WAY.
PUFF, PUFF, I'M COMING, I'M RIGHT HERE...
I'M SO PLEASED I WAS INVITED TO THIS CONFERENCE—IT'S WONDERFUL TO BE IN RIO!
FASHION
HOW ABOUT YOU, YUVAL?
YES, I'M REALLY GLAD, TOO, PUFF, PUFF!
I LIKE THEIR IDEA OF INVITING THE GENERAL PUBLIC TO THE CONFERENCE. IT'S USEFUL FOR EVERYONE TO UNDERSTAND MORE ABOUT OUR FORAGER ANCESTORS.
YES, AND THEY'VE GOT SOME EXCELLENT CONTRIBUTORS. I THINK IT'S GOING TO BE FASCINATING!
I HOPE YOU CATCH YOUR BREATH BY THE TIME WE GET THERE!
PUFF! PUFF!

I'M INTRIGUED ABOUT WHAT YOU'RE GOING TO SAY, YUVAL. WHAT SORT OF PAPER HAVE YOU PUT TOGETHER?

WELL, I'M PLANNING TO START BY SAYING THAT IF WE WANT TO UNDERSTAND OUR NATURE, HISTORY AND PSYCHOLOGY, WE NEED TO GET INSIDE THE HEADS OF OUR HUNTER-GATHERER ANCESTORS.

JUST TRYING TO PICTURE YOUR IDEA... HA, HA!

IN RECENT GENERATIONS, MORE AND MORE PEOPLE HAVE BEEN WORKING IN FACTORIES AND OFFICES.

TRUE, VERY TRUE!

BEFORE THAT, THERE WERE 10,000 YEARS WHEN MOST SAPIENS WERE FARMERS AND HERDERS.

BUT THAT'S THE BLINK OF AN EYE COMPARED TO THE TENS OF THOUSANDS OF YEARS WHEN OUR ANCESTORS HUNTED AND FORAGED.

120

THAT'S RIGHT. WE LIVED AS FORAGERS FOR NEARLY OUR ENTIRE HISTORY AS A SPECIES. AND EVOLUTIONARY PSYCHOLOGY SUGGESTS THAT A LOT OF THE SOCIAL AND PSYCHOLOGICAL CHARACTERISTICS WE HAVE TODAY WERE SHAPED IN THAT LONG PRE-AGRICULTURAL ERA.

SO OUR BRAINS ARE STILL ADAPTED FOR LIFE AS HUNTER-GATHERERS.

OUR EATING HABITS, OUR CONFLICTS AND OUR SEXUALITY ARE ALL THE RESULT OF OUR HUNTER-GATHERER MINDS GRAPPLING WITH A POST-INDUSTRIAL WORLD.

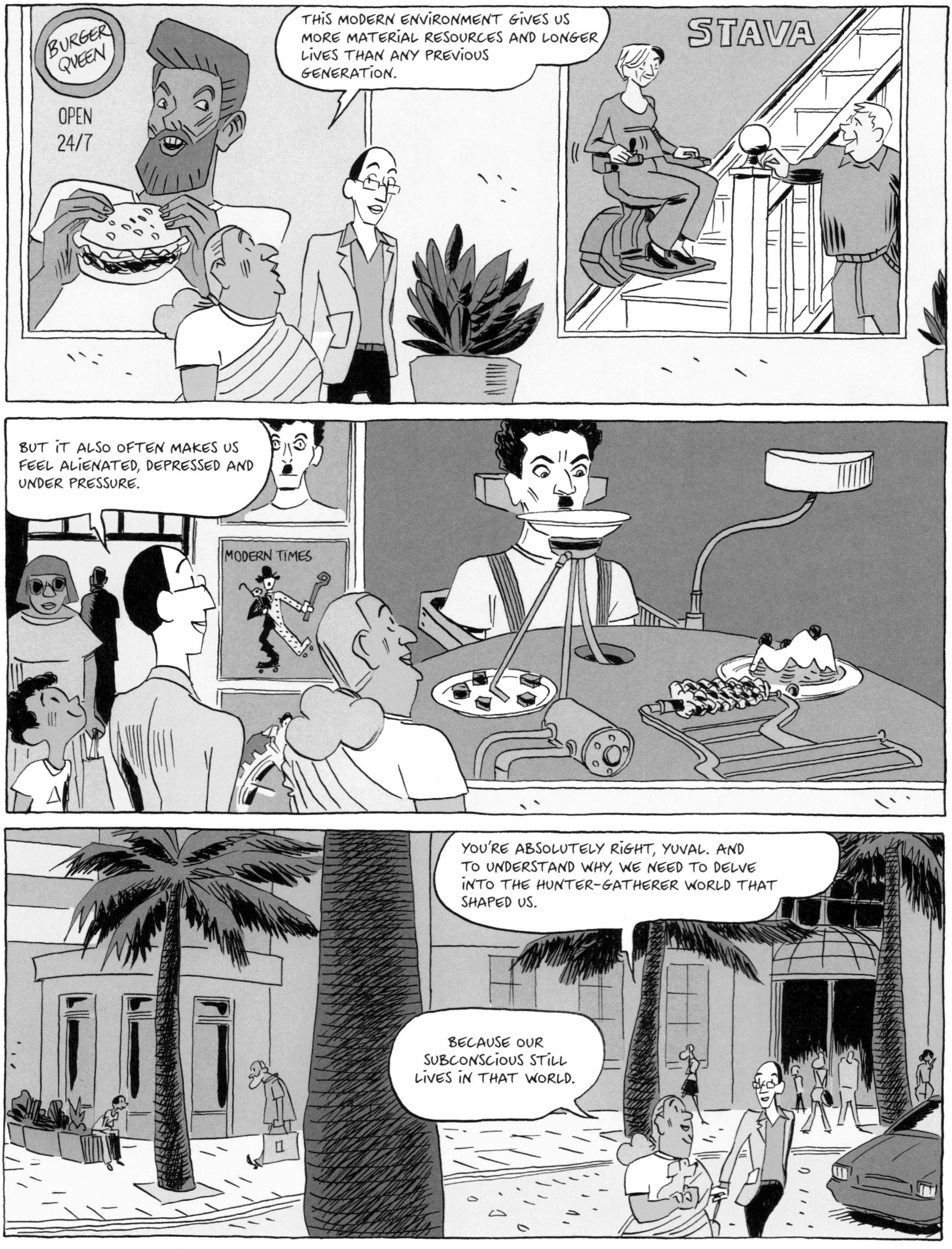
BURGER QUEEN
OPEN 24/7
THIS MODERN ENVIRONMENT GIVES US MORE MATERIAL RESOURCES AND LONGER LIVES THAN ANY PREVIOUS GENERATION.
STAVA
BUT IT ALSO OFTEN MAKES US FEEL ALIENATED, DEPRESSED AND UNDER PRESSURE.
MODERN TIMES
YOU'RE ABSOLUTELY RIGHT, YUVAL. AND TO UNDERSTAND WHY, WE NEED TO DELVE INTO THE HUNTER-GATHERER WORLD THAT SHAPED US.
BECAUSE OUR SUBCONSCIOUS STILL LIVES IN THAT WORLD.

WE'LL TAKE A TAXI TO THE TEATRO POPULAR IN NITERÓI.

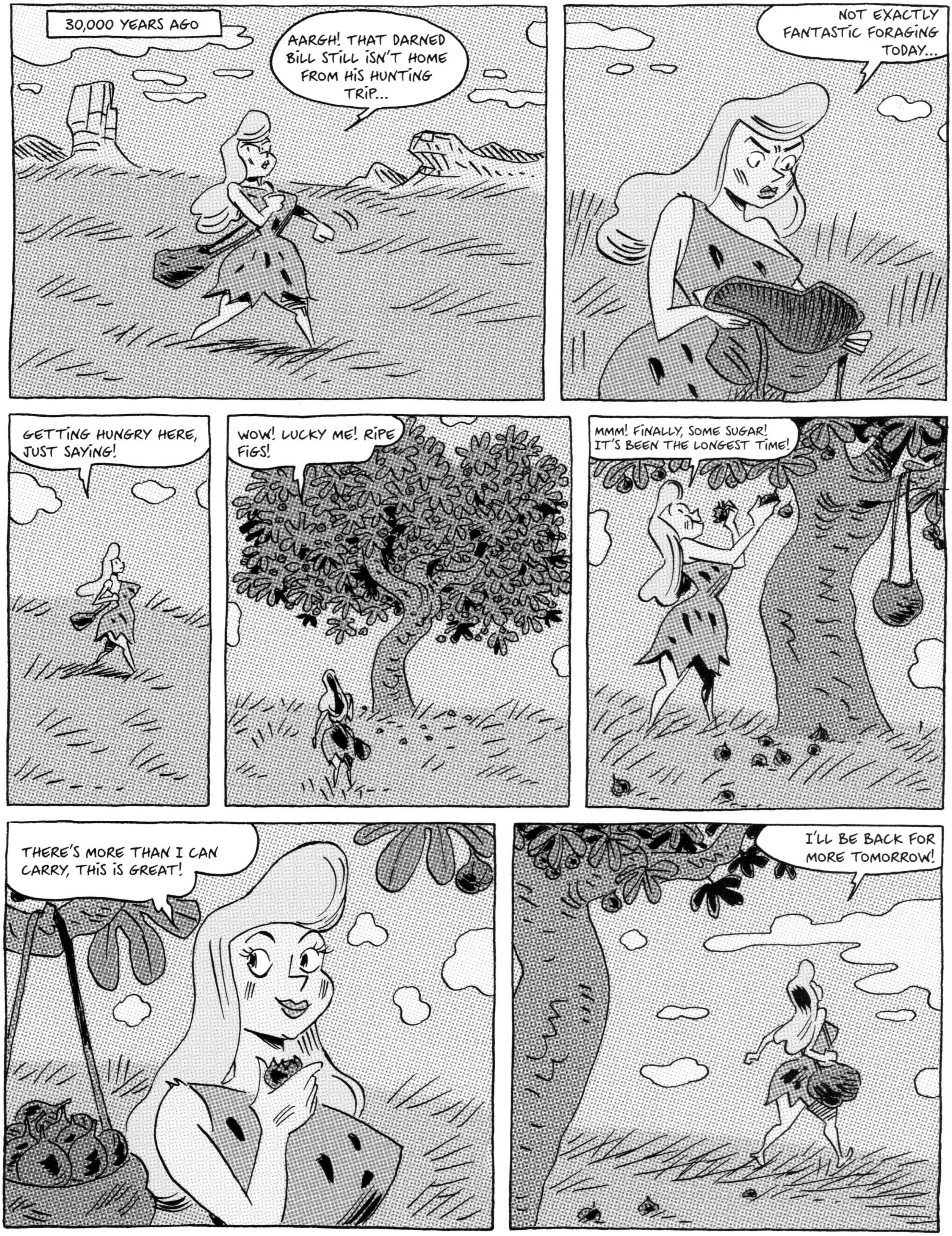
30,000 YEARS AGO
AARGH! THAT DARNED BILL STILL ISN'T HOME FROM HIS HUNTING TRIP...
NOT EXACTLY FANTASTIC FORAGING TODAY...
GETTING HUNGRY HERE, JUST SAYING!
WOW! LUCKY ME! RIPE FIGS!
MMM! FINALLY, SOME SUGAR! IT'S BEEN THE LONGEST TIME!
THERE'S MORE THAN I CAN CARRY, THIS IS GREAT!
I'LL BE BACK FOR MORE TOMORROW!

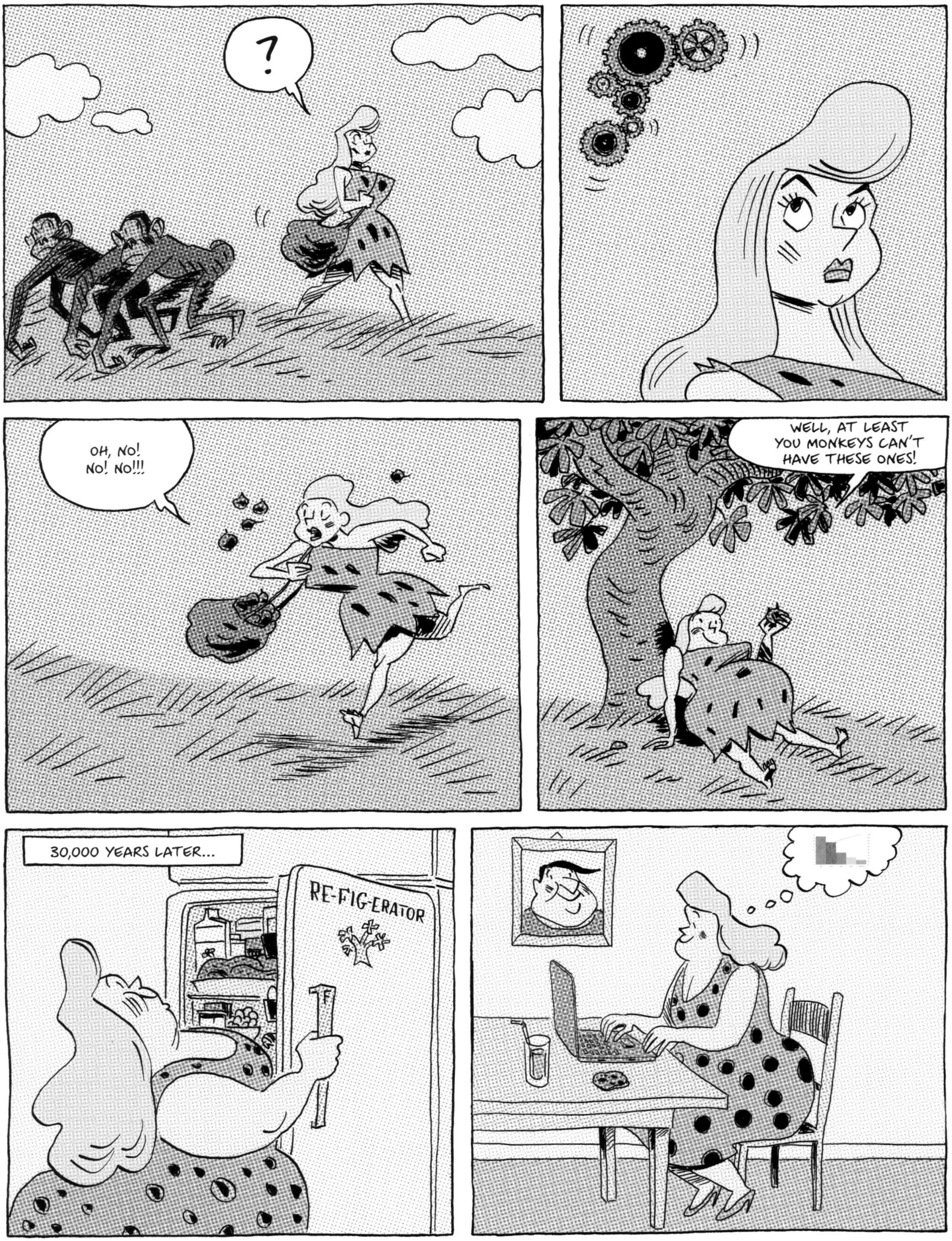
?
OH, NO! NO! NO!!!
WELL, AT LEAST YOU MONKEYS CAN'T HAVE THESE ONES!
30,000 YEARS LATER...
RE-FIG-ERATOR

I SEE ZOE'S GIVEN YOU THE COMICS BUG.

YES, IT'S REALLY GOOD, HERE, TAKE A LOOK.

HOW TRUE.

CORPORATIONS CAN TEMPT US WITH FOOD FULL OF SUGAR AND FAT RIGHT NOW BECAUSE OF TENDENCIES WE INHERITED FROM OUR HUNTER-GATHERER ANCESTORS.

THESE DAYS OBESITY AND DIABETES KILL MORE PEOPLE THAN WARS DO. BUT LIKING SUGAR AND FAT WAS PERFECTLY LOGICAL IN THE STONE AGE, AND IT'S NOW HARD-WIRED INTO OUR GENES. OUR DNA THINKS WE'RE STILL ON THE SAVANNA...

AH! WE'RE HERE ALREADY!

AM I RIGHT IN THINKING THAT THE "GORGING GENE" THEORY IS WIDELY ACCEPTED?

YES, YES.

BUT OTHER THEORIES ARE FAR MORE CONTENTIOUS. LIKE THE ONES ABOUT SEXUALITY AND FAMILY STRUCTURE... AHA! THERE'S THE THEATER!

I NOTICED IN THE PROGRAM THAT SEXUALITY AND FAMILY UNITS ARE MAJOR THEMES OF THE CONFERENCE.

I HOPE YOU'RE READY FOR SOME HEATED DEBATES!

HERE WE ARE!

AND ON TIME!

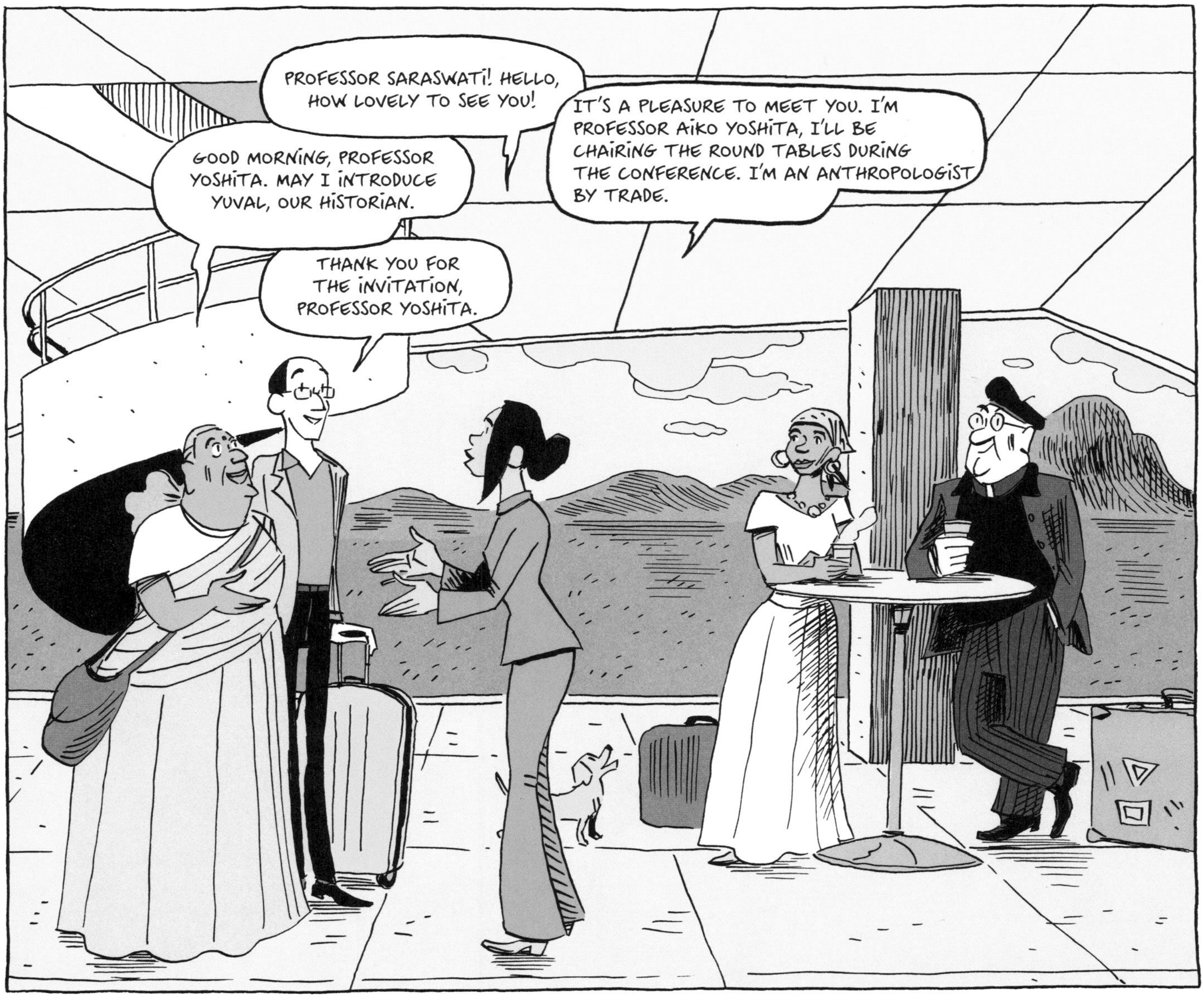
PROFESSOR SARASWATI! HELLO, HOW LOVELY TO SEE YOU!
GOOD MORNING, PROFESSOR YOSHITA. MAY I INTRODUCE YUVAL, OUR HISTORIAN.
THANK YOU FOR THE INVITATION, PROFESSOR YOSHITA.
IT'S A PLEASURE TO MEET YOU. I'M PROFESSOR AIKO YOSHITA, I'LL BE CHAIRING THE ROUND TABLES DURING THE CONFERENCE. I'M AN ANTHROPOLOGIST BY TRADE.

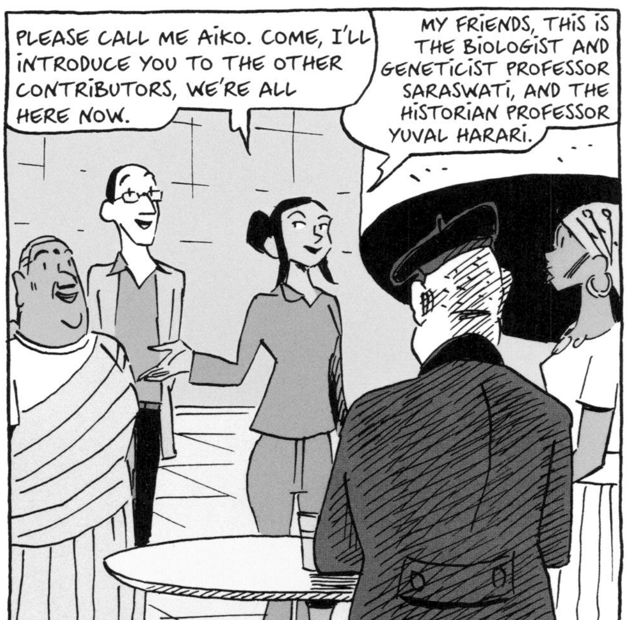
PLEASE CALL ME AIKO. COME, I'LL INTRODUCE YOU TO THE OTHER CONTRIBUTORS, WE'RE ALL HERE NOW.
MY FRIENDS, THIS IS THE BIOLOGIST AND GENETICIST PROFESSOR SARASWATI, AND THE HISTORIAN PROFESSOR YUVAL HARARI.

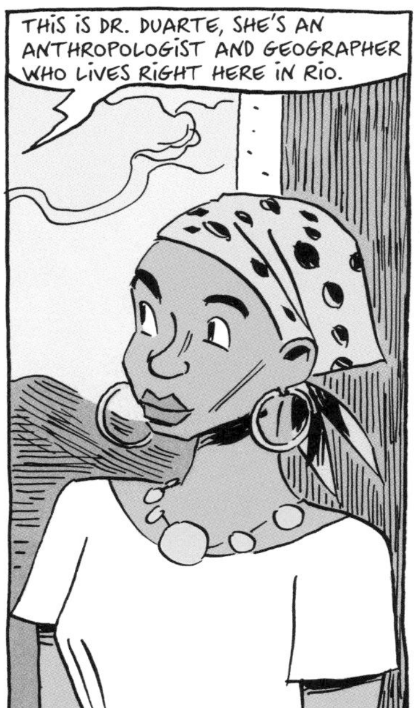
THIS IS DR. DUARTE, SHE'S AN ANTHROPOLOGIST AND GEOGRAPHER WHO LIVES RIGHT HERE IN RIO.

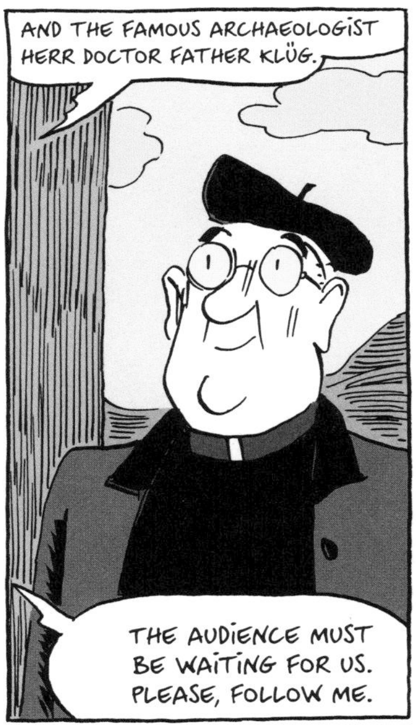
AND THE FAMOUS ARCHAEOLOGIST HERR DOCTOR FATHER KLÜG.
THE AUDIENCE MUST BE WAITING FOR US. PLEASE, FOLLOW ME.

LADIES AND GENTLEMEN, THANK YOU ALL FOR JOINING US FOR THIS ROUND TABLE. OVER THE NEXT HOUR WE'LL BE LOOKING AT HOW OUR ANCESTORS HOMO SAPIENS LIVED AFTER THE COGNITIVE REVOLUTION 70,000 AGO BUT BEFORE THE BEGINNING OF THE AGRICULTURAL REVOLUTION 12,000 YEARS AGO.

WHAT WAS LIFE LIKE IN THE STONE AGE?

WE'LL ALSO CONSIDER HOW WE KNOW THE THINGS THAT WE THINK WE KNOW. HOW CAN WE KNOW IF WE FOUGHT WARS 30,000 YEARS AGO? HOW DO WE ESTABLISH WHETHER WE HAD RELIGIOUS BELIEFS?

BUT TO START WITH, LET'S EXPLORE THE SUBJECT OF FAMILIES.

WHAT DID A TYPICAL STONE-AGE FAMILY LOOK LIKE, DOCTOR DUARTE?

EXCUSE ME, BUT WHY WOULD YOU START YOUR CONFERENCE WITH SUCH A POINTLESS QUESTION!

HONEY, PLEASE, DON'T MAKE A SCENE.

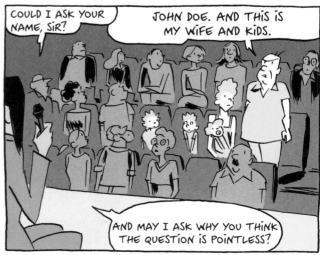

COULD I ASK YOUR NAME, SIR?

JOHN DOE. AND THIS IS MY WIFE AND KIDS.

AND MAY I ASK WHY YOU THINK THE QUESTION IS POINTLESS?

BECAUSE—JEEZ!—THE ANSWER'S OBVIOUS! HUMANS HAVE ALWAYS LIVED IN A FAMILY WITH A MOM, A DAD AND THEIR KIDS!

I THINK WE MIGHT HAVE TO SPEAK UP SOONER THAN WE WERE THINKING.

HMM... I'M AFRAID SO...

THAT'S BY NO MEANS A CERTAINTY. IN THE UNITED STATES TODAY, ONLY HALF OF ALL CHILDREN GROW UP IN A FAMILY LIKE THE ONE YOU JUST DESCRIBED. AND HERE IN BRAZIL, ONLY 50% OF HOUSEHOLDS ARE MADE UP OF A HETEROSEXUAL COUPLE AND THEIR BIOLOGICAL CHILDREN.

FAMILIES CAME IN ALL SHAPES AND SIZES IN PREVIOUS ERAS TOO...

I LIVED IN AN ALL-MALE COMMUNITY FOR MANY YEARS, YOU KNOW. THERE WERE NO WOMEN AT ALL, AND WE TOOK TURNS DOING THE COOKING, THE DISHES, THE HOUSEWORK...

A HOMOSEXUAL HIPPIE COMMUNE?!?

ACTUALLY, A BENEDICTINE MONASTERY IN BAVARIA. THERE WERE ABOUT 30 OF US BROTHERS, ONE FATHER AND NO MOTHER.

AND THE HOLY BIBLE IS FULL OF OTHER EXAMPLES. ABRAHAM'S WIFE, SARAH, SUGGESTED HE SHOULD SLEEP WITH THEIR EGYPTIAN SERVANT GIRL. AND JACOB HAD TWO WIVES AND TWO CONCUBINES—BETWEEN THEM THEY GAVE BIRTH TO 12 SONS AND WHO KNOWS HOW MANY DAUGHTERS!

SUSIE, WHERE ARE YOU?

HERE!

KING SOLOMON IS SAID TO HAVE HAD A THOUSAND WIVES...

OH COME ON! THESE OLD TESTAMENT GUYS WERE KIND OF EXTREME!

A THOUSAND WIVES IS... EXTREME. BUT A SINGLE SPOUSE FOR LIFE IS ALSO FAR FROM THE NORM FOR SAPIENS. SOME PEOPLE HAVE ONE PARTNER THEIR WHOLE LIVES, BUT OTHERS HAVE SEVERAL PARTNERS, AND THEN THERE ARE PEOPLE WHO STAY SINGLE...

LOOK AT THE ACTRESS ELIZABETH TAYLOR, FOR EXAMPLE. SHE WAS MARRIED EIGHT TIMES!

YES, BUT ONLY TO SEVEN DIFFERENT MEN BECAUSE I MARRIED RICHARD TWICE, HA! HA!

IN SOME COUNTRIES, SUCH AS SAUDI ARABIA, IT'S LEGAL FOR ONE MAN TO HAVE SEVERAL WIVES AT THE SAME TIME.

BUT IN OTHER COUNTRIES TWO WOMEN CAN MARRY EACH OTHER, OR TWO MEN. FOR EXAMPLE, XAVIER BETTEL, THE PRIME MINISTER OF LUXEMBOURG, IS MARRIED TO ANOTHER MAN, GAUTHIER DESTENAY.

PAH! BUT THAT'S A TINY COUNTRY!

SOME PEOPLE HAVE ONE CHILD, OTHERS HAVE TEN, AND STILL OTHERS ARE HAPPY AS THEY ARE, WITH NO CHILDREN. MOHAMMED BIN AWAD BIN LADEN, FATHER OF OSAMA BIN LADEN, HAD MORE THAN 50 CHILDREN.

THE POPE, WHO'S THE HEAD OF THE ROMAN CATHOLIC CHURCH, HAS NO CHILDREN. AND HE SEEMS VERY HAPPY WITH THAT.

SOME CHILDREN ARE RAISED BY A SINGLE MOTHER, OR A SINGLE FATHER, OR MAYBE BY GRANDPARENTS. SOME CHILDREN ARE ADOPTED. SOME CHILDREN HAVE TWO FATHERS, OTHERS HAVE TWO MOTHERS.

SOMETIMES, PARENTS SEPARATE AND THEN FIND NEW PARTNERS, SO A CHILD CAN HAVE A MOTHER AND A FATHER BUT ALSO A STEPFATHER AND A STEPMOTHER.

IN SOME FAMILIES DOZENS OF AUNTS, UNCLES, COUSINS AND GRANDPARENTS ALL LIVE TOGETHER... SO YOU MIGHT SHARE A BEDROOM WITH YOUR COUSIN INSTEAD OF YOUR BROTHER, AND YOUR UNCLE OR YOUR GRANDMOTHER MIGHT MAKE YOUR BREAKFAST EVERY DAY INSTEAD OF YOUR PARENTS.

THERE ARE SO MANY VARIATIONS!

BUT THERE'S ONLY ONE THAT'S NATURAL! ALL THE OTHERS ARE... WELL, THEY'RE NOT NATURAL!

NOT NATURAL? MAYBE OUR PRIMATE COUSINS CAN TEACH US SOMETHING ABOUT NATURAL FAMILIES...

131

GIBBONS USUALLY LIVE IN COUPLES. WHEN A MALE AND A FEMALE FORM A COUPLE, THEY TEND TO STAY TOGETHER FOR MANY YEARS. THEY LIVE ALONE TOGETHER, IN THEIR OWN PART OF THE FOREST, AND LOOK AFTER THEIR OWN YOUNG.

THERE! YOU SEE! IT'S NATURAL!

BUT WAIT... WITH GORILLAS, IT'S FAR MORE COMMON FOR A MALE TO LIVE WITH A HAREM OF MANY FEMALES AND ALL OF THEIR YOUNG. EACH BABY GORILLA HAS A DIFFERENT MOTHER, BUT THEY ALL HAVE THE SAME FATHER.

ORANGUTANS LIKE THEIR SOLITUDE, THEY ENJOY SITTING ALONE PEACEFULLY, PERHAPS JUST WATCHING THE SUNSET. ORANGUTAN MOTHERS ARE ALMOST ALWAYS SINGLE PARENTS. THEY RAISE THEIR CHILDREN SINGLE-HANDED. AND WHEN THE YOUNG MATURE, THEY LEAVE TO GO AND LIVE ON THEIR OWN. THAT'S THE WAY THEY LIKE IT.

CHIMPANZEES ARE THE EXACT OPPOSITE OF ORANGUTANS. THEY LIVE IN NOISY COMMUNITIES OF LOTS OF MALES AND FEMALES TOGETHER. THEY DON'T FORM LASTING COUPLES. YOUNG CHIMPS STAY VERY CLOSE TO THEIR MOTHERS, BUT USUALLY DON'T EVEN KNOW WHO THEIR FATHER IS. THE WORD "FATHER" PROBABLY WOULDN'T MEAN MUCH TO THEM.

IN ONE TYPE OF CHIMPANZEE, KNOWN AS THE COMMON CHIMPANZEE, THE MALES HANG OUT TOGETHER, AND THE MOST POWERFUL MALE IS THE LEADER OF THE WHOLE GROUP.

BUT WITH ANOTHER TYPE OF CHIMPANZEES, BONOBOS, THE FEMALES FORM VERY CLOSE FRIENDSHIPS—SEX INCLUDED! THEY HELP EACH OTHER RAISE THEIR YOUNG, AND THEY'RE THE ONES WHO MAKE DECISIONS, NOT THE ADULT MALES. LITTLE BONOBO GIRLS DON'T DREAM ABOUT MARRYING A PRINCE. MOST OF THEM WOULD PREFER A COOL GIRLFRIEND INSTEAD.

OK, SO I DIDN'T KNOW ALL THAT... BUT YOU'RE TALKING ABOUT FREAKING APES! I WANNA KNOW HOW *HUMAN* FAMILIES LIVED IN THE STONE AGE!

THAT'S EXACTLY THE QUESTION WE'RE TRYING TO ANSWER... SO, I HOPE YOU'LL AGREE, IT'S NOT POINTLESS, AFTER ALL.

IF I COULD JUMP IN HERE, SOME ANTHROPOLOGISTS THINK THAT THE MOST COMMON ARRANGEMENT DURING THE STONE AGE WAS TO LIVE IN LARGE FAMILY COMMUNITIES, RATHER LIKE CHIMPANZEES.

ALL THE ADULTS CONTRIBUTED TO RAISING ALL THE CHILDREN. THERE WAS NO CLEAR DISTINCTION BETWEEN FAMILY UNITS. THERE WERE NO MARRIAGES... AND NO DIVORCE LAWYERS EITHER!

IF YOU LIKED SOMEONE, YOU JUST MOVED YOUR PILLOW OF DRIED GRASS NEXT TO THEIRS. IF YOU GOT BORED OF EACH OTHER, YOU MOVED YOUR PILLOW AWAY AGAIN...

WELL, IF YOU HAD A PILLOW, THAT IS!

BUT WHAT ABOUT THE KIDS, THEN?

CHILDREN KNEW WHO THEIR MOTHER WAS, OBVIOUSLY. BECAUSE THEIR MOTHERS GAVE BIRTH TO THEM AND LOOKED AFTER THEM FOR YEARS.

BUT WE CAN'T BE SURE THAT THEY KNEW WHO THEIR FATHER WAS. THE MEN PROBABLY HELPED RAISE ALL THE CHILDREN BY BRINGING FOOD, PROTECTING THEM, TEACHING THEM TO CLIMB TREES OR MAKE STONE TOOLS...

CHILDREN COULD HAVE HAD CLOSE TIES WITH SEVERAL ADULTS, AND NO ONE FELT THE NEED TO DEFINE EXACTLY WHO WAS A FATHER OR AN UNCLE OR AN UNRELATED NEIGHBOR...

SO YOU SEE, MR. DOE, IT WAS A LITTLE LIKE THE SET-UP WITH OUR COUSINS THE CHIMPANZEES.

ANTHROPOLOGISTS HAVE ACTUALLY OBSERVED SOME PRESENT-DAY HUMAN CULTURES THAT PRACTICE COLLECTIVE FATHERHOOD.

THE BARI INDIANS, FOR EXAMPLE. THEY BELIEVE THAT A CHILD ISN'T BORN FROM THE SPERM OF JUST ONE MAN, BUT FROM THE ACCUMULATION OF SPERM FROM DIFFERENT MEN IN A WOMAN'S WOMB.

SO A GOOD MOTHER MAKES A POINT OF HAVING SEX WITH SEVERAL MEN, ESPECIALLY WHEN SHE'S PREGNANT, SO THAT HER CHILD INHERITS GOOD QUALITIES—AND ENJOYS PATERNAL CARE—FROM NOT JUST THE BEST HUNTER, BUT ALSO THE BEST STORYTELLER, THE BRAVEST WARRIOR AND THE MOST CONSIDERATE LOVER.

THIS IS TOTALLY CRAZY! COME ON, HONEY, WE'RE LEAVING! THESE SCIENTISTS ARE TREATING US LIKE IDIOTS!

WE'RE VERY SORRY WE UPSET YOU, MR. DOE. BUT DON'T FORGET THAT, BEFORE OUR MODERN BOOM IN EMBRYOLOGICAL STUDIES, THERE WAS NO WAY OF PROVING THAT A BABY WAS ALWAYS PRODUCED BY A SINGLE FATHER'S SPERM.

PICTURE YOURSELF IN AN ANCIENT HUNTER-GATHERER TRIBE. HOW COULD YOU PROVE TO PEOPLE THAT EVERY CHILD HAD JUST ONE FATHER, AND NEVER THREE!

HMMMM...

YOU KNOW, WE CATHOLICS BELIEVE THAT OUR FATHER IS THREE ...

IF EARLY HUMANS REALLY DID LIVE IN LARGE COMMUNAL FAMILIES LIKE CHIMPANZEES, THAT WOULD EXPLAIN A LOT...

IT WOULD EXPLAIN THE FREQUENT INFIDELITIES IN MODERN MARRIAGES AND THE HIGH DIVORCE RATE...

MAYBE THAT'S WHY A NUCLEAR FAMILY AND MONOGAMY AREN'T THAT EASY FOR EVERYONE...

BUT OBVIOUSLY THAT'S JUST A THEORY AND IT DOESN'T EXPLAIN EVERYTHING.

IN MANY CULTURES, NUCLEAR FAMILIES AND MONOGAMOUS MARRIAGES ARE THE NORM.

HUMANS ARE VERY COMPLICATED, AND WE DON'T KNOW ENOUGH ABOUT THE STONE AGE TO REACH A DEFINITIVE CONCLUSION.

THEORIES ARE IMPORTANT TO SCIENTISTS, BUT AT THE END OF THE DAY, IT'S ALL ABOUT HAVING PROOF...

TELL ME, FATHER KLÜG, AS AN ARCHAEOLOGIST... WHAT ARCHAEOLOGICAL CLUES DO WE HAVE ABOUT DAILY LIFE IN THE STONE AGE?

WHEN WE DISCUSS ANCIENT FAMILY STRUCTURES, PEOPLE OFTEN GET VERY... UM, SENSITIVE, YOU COULD EVEN SAY IRATE...

THE TRUTH IS, THE WHOLE DEBATE IS BASED ON VERY FEW CLUES...

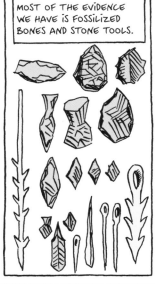

MOST OF THE EVIDENCE WE HAVE IS FOSSILIZED BONES AND STONE TOOLS.

THIS GIVES US THE INACCURATE IMPRESSION THAT OUR PREAGRICULTURAL ANCESTORS LIVED IN A STONE AGE. IT WOULD BE MORE ACCURATE TO CALL IT THE WOOD AGE.

WOOD

LEATHER

BAMBOO

ANY ATTEMPT TO RECONSTRUCT THE LIVES OF ANCIENT HUNTER-GATHERERS BASED ON SURVIVING ARTIFACTS IS EXTREMELY PROBLEMATIC. PARTICULARLY BECAUSE, UNLIKE THEIR DESCENDANTS, THEY USED VERY FEW ARTIFACTS IN THE FIRST PLACE!

NOWADAYS, A TYPICAL MEMBER OF OUR AFFLUENT SOCIETY WILL USE SEVERAL MILLION ARTIFACTS OVER THE COURSE OF HIS OR HER LIFE. THIS COULD BE FOR EATING...

PLAYING...

ROMANCE...

PRAYING...

IT'S ONLY WHEN WE MOVE TO A NEW APARTMENT THAT WE REALIZE JUST HOW MUCH STUFF WE HAVE!

BUT BACK TO OUR FORAGERS—THEY WERE NOMADIC. THEY RELOCATED EVERY MONTH, SOMETIMES EVEN EVERY DAY. THEY CARRIED ALL THEIR BELONGINGS ON THEIR BACKS. THERE WERE NO MOVERS, NO WAGONS, NOT EVEN PACK ANIMALS. SO THEY HAD TO MAKE DO WITH THE ESSENTIALS.

COULD WE CONCLUDE, THEN, FATHER KLÜG, THAT MOST OF THEIR RELIGIOUS, EMOTIONAL AND SOCIAL LIVES WERE CONDUCTED WITHOUT ARTIFACTS?

ABSOLUTELY. IN 100,000 YEARS' TIME, AN ARCHAEOLOGIST COULD FORM A REASONABLE IDEA OF ISLAMIC BELIEFS AND PRACTICES BASED ON THE MYRIAD OBJECTS UNEARTHED FROM EXCAVATED MOSQUE SITES...

BUT WE'RE AT A LOSS TRYING TO UNDERSTAND HUNTER-GATHERER BELIEFS AND RITUALS.

ONE WAY TO REMEDY THIS PROBLEM IS TO OBSERVE MODERN-DAY FORAGER SOCIETIES, LIKE ABORIGINAL TRIBES IN AUSTRALIA, OR HUNTER-GATHERERS IN THE AMAZON. THEY CAN BE STUDIED FIRST-HAND.

YES, BUT WE MUST BE EXTREMELY CAREFUL NOT TO JUMP TO CONCLUSIONS.

EXCUSE ME... I COME FROM AN AMAZONIAN TRIBAL COMMUNITY. LOTS OF US DON'T LIVE LIKE HUNTER-GATHERERS NOWADAYS. SOME ARE FARMERS, OTHERS WORK IN CITIES.

TWO HUNDRED YEARS AGO, MY PEOPLE WERE FORAGERS, YES. BUT NO WAY DOES THAT MEAN THEY LIVED EXACTLY THE SAME WAY AS THE PEOPLE WHO DID CAVE PAINTINGS 20,000 YEARS AGO.

THANK YOU SO MUCH. THAT'S A VERY VALID POINT.

YOU'RE EXACTLY RIGHT. FIRSTLY, BECAUSE ALL FORAGER SOCIETIES THAT STILL EXIST TODAY HAVE BEEN INFLUENCED BY NEIGHBORING AGRICULTURAL AND INDUSTRIAL SOCIETIES.

SECONDLY, MODERN FORAGER SOCIETIES HAVE MOSTLY SURVIVED IN REGIONS WITH DIFFICULT CLIMATIC CONDITIONS, AND INHOSPITABLE TERRAIN THAT DOESN'T LEND ITSELF TO AGRICULTURE.

AS AN EXAMPLE, WE TEND TO ASSOCIATE AUSTRALIA'S ABORIGINE FORAGERS WITH A DESERT ENVIRONMENT. BUT WHEN EUROPEANS FIRST ARRIVED, MOST ABORIGINES LIVED IN FERTILE LAND AROUND WHAT'S NOW SYDNEY AND MELBOURNE.

SOCIETIES THAT HAVE ADAPTED TO EXTREME CONDITIONS IN PLACES LIKE THE KALAHARI DESERT IN SOUTHERN AFRICA MAY WELL BE MISLEADING MODELS FOR STONE-AGE SOCIETIES IN MORE FERTILE REGIONS.

FOR ONE THING, PRESENT-DAY POPULATION DENSITY IN SOMEWHERE LIKE THE KALAHARI DESERT IS MUCH LOWER THAN IT WAS IN THE YANGTZE VALLEY IN THE STONE AGE. AND THIS HAS FAR-REACHING IMPLICATIONS FOR KEY QUESTIONS ABOUT THE SIZE AND STRUCTURE OF HUMAN COMMUNITIES, AND HOW THEY INTERACTED WITH OTHER COMMUNITIES.

THAT'S RIGHT. AND EVEN MORE IMPORTANTLY, THE MOST OBVIOUS CHARACTERISTIC OF HUNTER-GATHERER SOCIETIES IS HOW DIVERSE THEY ARE! WE MIGHT THINK IF YOU KNOW ONE FORAGER, YOU KNOW THEM ALL... BUT THE TRUTH IS, FORAGER SOCIETIES VARY A WHOLE LOT, NOT JUST FROM ONE PART OF THE WORLD TO ANOTHER, BUT EVEN WITHIN THE SAME REGION!

THE AMAZING VARIETY THAT THE FIRST EUROPEAN COLONIZERS FOUND AMONG ABORIGINAL PEOPLES IN AUSTRALIA IS A GREAT EXAMPLE. JUST BEFORE THE BRITISH CONQUEST, THERE WERE BETWEEN 300,000 AND 700,000 HUNTER-GATHERERS LIVING ON THE CONTINENT IN 200 TO 600 TRIBES. AND EACH TRIBE WAS MADE UP OF SEVERAL SMALLER BANDS.

EACH TRIBE HAD ITS OWN LANGUAGE, NORMS AND CUSTOMS, WHICH COULD BE VERY DIFFERENT FROM THOSE IN THE NEIGHBORING TRIBE. AROUND WHAT IS NOW ADELAIDE IN SOUTHERN AUSTRALIA, THERE WERE SEVERAL WHAT WE CALL "PATRILINEAL CLANS." THAT MEANS THEY ESTABLISHED DESCENT FROM THE FATHER'S SIDE. THEIR CLANS BONDED INTO STRICTLY TERRITORIAL TRIBES.

MEANWHILE, SOME TRIBES IN NORTHERN AUSTRALIA FOCUSED MORE ON AN INDIVIDUAL'S MATERNAL LINE OF DESCENT, AND PEOPLE'S TRIBAL IDENTITY DEPENDED ON THEIR TOTEM RATHER THAN THEIR TERRITORY.

IT STANDS TO REASON, THEN, THAT THERE WAS EQUALLY IMPRESSIVE ETHNIC AND CULTURAL DIVERSITY AMONG ANCIENT HUNTER-GATHERERS, AND THE FIVE TO EIGHT MILLION FORAGERS WHO WALKED THE EARTH JUST BEFORE THE AGRICULTURAL REVOLUTION WERE DIVIDED INTO THOUSANDS OF SEPARATE TRIBES, WITH THOUSANDS OF DIFFERENT LANGUAGES AND CULTURES.

IN FACT, SURELY THAT'S ONE OF THE MAIN LEGACIES OF THE COGNITIVE REVOLUTION. WOULD YOU AGREE, YUVAL?

VERY MUCH SO! THANKS TO THEIR CAPACITY FOR FICTION, EVEN POPULATIONS WITH THE SAME GENETIC MAKE-UP WHO LIVED UNDER SIMILAR ECOLOGICAL CONDITIONS COULD CREATE VERY DIFFERENT IMAGINED REALITIES, THEREFORE ESTABLISHING DIFFERENT NORMS AND VALUES.

FOR EXAMPLE, WE HAVE EVERY REASON TO BELIEVE THAT A FORAGER BAND LIVING 30,000 YEARS AGO ON THE SPOT WHERE OXFORD UNIVERSITY NOW STANDS WOULD NOT HAVE SPOKEN THE SAME LANGUAGE AS ONE LIVING WHERE WE WOULD NOW FIND CAMBRIDGE UNIVERSITY.

CAMBRIDGE

OXFORD

LONDON

OXFORD

CAMBRIDGE

THESE BANDS COULD HAVE BEEN COMPLETELY DIFFERENT SOCIALLY AND POLITICALLY...

IN THEIR FAMILY MAKE-UP...

THEIR TABOOS...

AND, OF COURSE, IN THEIR RELIGIOUS BELIEFS.

IN OTHER WORDS, ANTHROPOLOGICAL OBSERVATIONS OF MODERN FORAGERS CAN HELP US UNDERSTAND SOME OPTIONS THAT WERE OPEN TO ANCIENT FORAGERS, BUT THE HORIZON OF POSSIBILITIES WAS MUCH WIDER THEN... AND SADLY, WE CAN'T SEE THAT MUCH OF IT NOW.

HEATED DEBATES ABOUT HOMO SAPIENS' "NATURAL WAY OF LIFE" ARE MISSING THE MAIN POINT.

EVER SINCE THE COGNITIVE REVOLUTION, THERE HASN'T BEEN JUST ONE NATURAL WAY OF LIFE FOR SAPIENS. INSTEAD, THERE ARE CULTURAL CHOICES TAKEN FROM A BEWILDERING SELECTION OF OPTIONS.

THANK YOU VERY MUCH, YUVAL. NOW, BEFORE WE BRING THIS VERY INTERESTING DISCUSSION TO A CLOSE, WE HAVE A SHORT FILM ABOUT SOME OF THE THINGS THAT WE KNOW ABOUT ANCIENT FORAGERS.

LIGHTS, PLEASE.

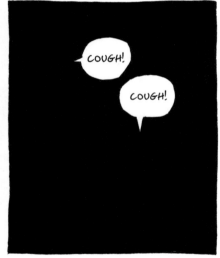

COUGH!

COUGH!

LIFE IN THE STONE AGE

WHAT DO WE REALLY KNOW ABOUT THE FORAGERS WHO LIVED IN THE ERAS BEFORE THE AGRICULTURAL REVOLUTION?

A FERTILE VALLEY BETWEEN 70,000 AND 12,000 YEARS AGO.

OF COURSE, THEIR TRIBES WERE ALL VERY DIFFERENT, SO CAN WE REALLY MAKE ANY GENERALIZATIONS?

IT SEEMS SAFE TO SAY THAT THE VAST MAJORITY OF PEOPLE LIVED IN SMALL BANDS OF SEVERAL DOZEN OR AT MOST A FEW HUNDRED INDIVIDUALS.

MEMBERS WITHIN A BAND KNEW EACH OTHER VERY INTIMATELY AND WERE SURROUNDED BY FRIENDS AND RELATIVES ALL THROUGH THEIR LIVES. LONELINESS AND PRIVACY WOULD HAVE BEEN RARE.

NEIGHBORING BANDS PROBABLY COMPETED FOR RESOURCES AND OCCASIONALLY MIGHT EVEN HAVE FOUGHT ONE ANOTHER.

BUT THEY WOULD HAVE HAD FRIENDLY CONTACT TOO. THEY EXCHANGED MEMBERS, SHARED INFORMATION, HUNTED TOGETHER AND TRADED RARE LUXURIES SUCH AS SEASHELLS.

SOME NEIGHBORING BANDS SOMETIMES JOINED FORCES AGAINST FOREIGNERS.

THIS SORT OF COOPERATION WAS AN IMPORTANT TRADEMARK OF HOMO SAPIENS, GIVING THEM A CRUCIAL EDGE OVER OTHER HUMAN SPECIES.

SOMETIMES RELATIONS WITH NEIGHBORING BANDS WERE TIGHT ENOUGH FOR ALL OF THEM TO CONSTITUTE A SINGLE TRIBE WITH A COMMON LANGUAGE, COMMON MYTHS AND COMMON NORMS AND VALUES. A TRIBE LIKE THIS MIGHT HAVE NUMBERED SEVERAL THOUSAND SAPIENS.

BUT WE MUST BE CAREFUL NOT TO OVERESTIMATE THE IMPORTANCE OF SUCH EXTERNAL RELATIONS. TRIBES CAME TOGETHER ONLY RARELY, SAY FOR SOME ANNUAL FESTIVAL. MOST OF THE TIME, SMALL BANDS LIVED THEIR LIVES IN COMPLETE ISOLATION AND PERFECTLY INDEPENDENTLY.

THE TRIBE WAS NOT A PERMANENT POLITICAL FRAMEWORK, AND EVEN IF IT HAD SEASONAL MEETING PLACES, THERE WERE NO PERMANENT TOWNS OR INSTITUTIONS.

THE AVERAGE PERSON COULD GO MANY MONTHS WITHOUT SEEING OR HEARING A HUMAN FROM ANOTHER BAND, AND MIGHT SEE NO MORE THAN A FEW THOUSAND HUMANS IN A LIFETIME.

THESE BANDS OF SAPIENS WERE THINLY SPREAD OVER VAST TERRITORIES. BEFORE THE AGRICULTURAL REVOLUTION, THE HUMAN POPULATION OF THE ENTIRE PLANET WAS SMALLER THAN THE POPULATION OF A BIG CITY LIKE TOKYO OR CAIRO TODAY.

MOST SAPIENS BANDS LIVED ON THE MOVE, ROAMING FROM PLACE TO PLACE IN SEARCH OF FOOD.

THEIR MOVEMENTS WERE INFLUENCED BY THE CHANGING SEASONS...

ANNUAL ANIMAL MIGRATIONS...

AND THE GROWTH CYCLE OF PLANTS.

BANDS USUALLY CAME AND WENT ACROSS THE SAME HOME TERRITORY—AN AREA OF BETWEEN A FEW DOZEN AND HUNDREDS OF SQUARE MILES.

BANDS WOULD VENTURE OUTSIDE THEIR TURF ONLY OCCASIONALLY TO EXPLORE NEW TERRITORIES. PERHAPS SPURRED ON BY A NATURAL DISASTER...

VIOLENT CONFLICT...

OR THE INITIATIVE OF A CHARISMATIC LEADER.

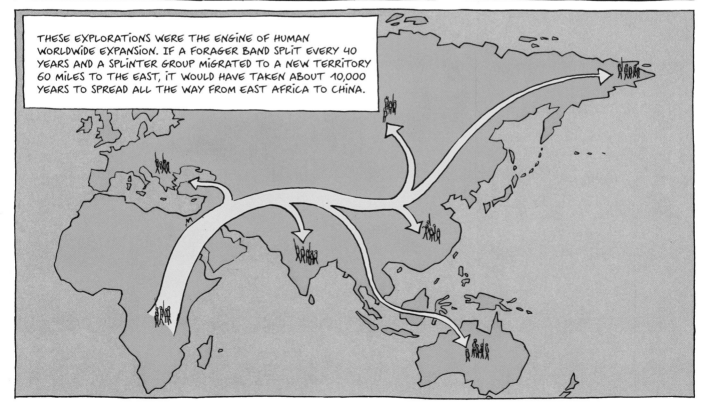

THESE EXPLORATIONS WERE THE ENGINE OF HUMAN WORLDWIDE EXPANSION. IF A FORAGER BAND SPLIT EVERY 40 YEARS AND A SPLINTER GROUP MIGRATED TO A NEW TERRITORY 60 MILES TO THE EAST, IT WOULD HAVE TAKEN ABOUT 10,000 YEARS TO SPREAD ALL THE WAY FROM EAST AFRICA TO CHINA.

IN SOME EXCEPTIONAL CASES WHEN FOOD WAS PARTICULARLY PLENTIFUL, BANDS SETTLED IN PERMANENT CAMPS.

TECHNIQUES FOR DRYING AND SMOKING FOOD ALSO ALLOWED THEM TO STAY IN ONE PLACE FOR LONGER PERIODS.

AND IN ARCTIC REGIONS THEY COULD PRESERVE FOOD BY FREEZING IT.

THE FIRST PERMANENT SETTLEMENTS IN HISTORY APPEARED LONG BEFORE THE AGRICULTURAL REVOLUTION, ALONG COASTS AND RIVERS RICH IN FISH, SEAFOOD AND WATERFOWL.

FISHING VILLAGES MIGHT HAVE APPEARED ON THE COASTS OF INDONESIAN ISLANDS SOME 50,000 YEARS AGO, IF NOT EARLIER.

AND THESE MAY HAVE BEEN LAUNCHPADS FOR HOMO SAPIENS' FIRST TRANSOCEANIC ENTERPRISE: COLONIZING AUSTRALIA.

IN MOST HABITATS, BANDS OF SAPIENS HAD MANY DIFFERENT SOURCES OF FOOD. THEY SCROUNGED FOR TERMITES...

PICKED BERRIES...

DUG FOR ROOTS...

STALKED SMALL GAME...

AND HUNTED LARGER GAME.

DESPITE THE POPULAR IMAGE OF "MAN THE HUNTER," FORAGING WAS STILL SAPIENS' MAIN ACTIVITY. IT PROVIDED MOST OF THEIR CALORIES AS WELL AS RAW MATERIALS SUCH AS FLINT, WOOD AND BAMBOO.

NO MATTER HOW THEY GOT THEIR FOOD, THE MOST IMPORTANT THING ALL SAPIENS NEEDED WAS KNOWLEDGE...

KNOWLEDGE OF GEOGRAPHY. THEY NEEDED INFORMATION ABOUT THEIR TERRITORY.

KNOWLEDGE OF ANIMALS. TO HUNT, THEY NEEDED TO BE FAMILIAR WITH THE HABITS OF EVERY ANIMAL.

KNOWLEDGE OF PLANTS. TO FIND FOOD, THEY NEEDED TO KNOW WHERE TO FIND PLANTS, BUT ALSO WHICH FOODS WERE NOURISHING, WHICH MADE YOU SICK AND WHICH HAD HEALING PROPERTIES.

KNOWLEDGE OF THE WEATHER. THEY NEEDED TO LEARN THE CYCLE OF THE SEASONS AND KNOW HOW TO RECOGNIZE SIGNS OF AN IMPENDING THUNDERSTORM OR A DRY SPELL.

THEY STUDIED EVERY STREAM, EVERY WALNUT TREE, EVERY BEAR CAVE AND EVERY FLINT-STONE DEPOSIT NEAR THEIR SETTLEMENT.

EACH INDIVIDUAL NEEDED TO KNOW HOW TO MAKE A STONE KNIFE, MEND A TORN CLOAK, SET A RABBIT TRAP AND COPE WITH AVALANCHES, SNAKEBITES AND HUNGRY LIONS. IT TOOK YEARS OF APPRENTICESHIP AND PRACTICE TO MASTER EACH OF THESE SKILLS.

FORAGERS MASTERED NOT ONLY THE WORLD AROUND THEM WITH ITS ANIMALS, PLANTS AND OBJECTS, BUT ALSO THE INNER WORLD OF THEIR OWN BODIES AND SENSES.

THEY LISTENED TO THE SLIGHTEST MOVEMENT IN THE GRASS IN CASE IT MEANT A SNAKE WAS LURKING THERE.

THEY CAREFULLY OBSERVED THE FOLIAGE OF TREES IN THE HOPES OF SPOTTING FRUIT...

A BEEHIVE...

OR A BIRD'S NEST.

THEY MOVED WITH THE MINIMUM OF EFFORT OR NOISE, AND KNEW THE MOST AGILE, EFFICIENT WAY TO SIT, RUN AND WALK. CONSTANT, VARIED USE OF THEIR BODIES MADE THEM AS FIT AS MARATHON RUNNERS.

ANCIENT FORAGERS ENJOYED PHYSICAL DEXTERITY THAT PEOPLE TODAY CANNOT ACHIEVE EVEN AFTER YEARS OF PRACTICING YOGA OR TAI CHI.

YUVAL, WOULD YOU LIKE TO SAY A FEW WORDS TO SUM UP WHAT WE'VE COVERED SO FAR?

THE END

YES, THANK YOU, AIKO. I'D LIKE TO COME BACK TO SOMETHING THAT I THINK IS VERY IMPORTANT.

AS HAS BEEN CLEARLY EXPLAINED, SURVIVAL IN THE STONE AGE REQUIRED EVERY INDIVIDUAL TO HAVE SUPERB PHYSICAL AND MENTAL ABILITIES.

IN FACT, THERE'S EVIDENCE THAT THE AVERAGE SAPIENS BRAIN HAS ACTUALLY DECREASED SINCE OUR FORAGING DAYS!

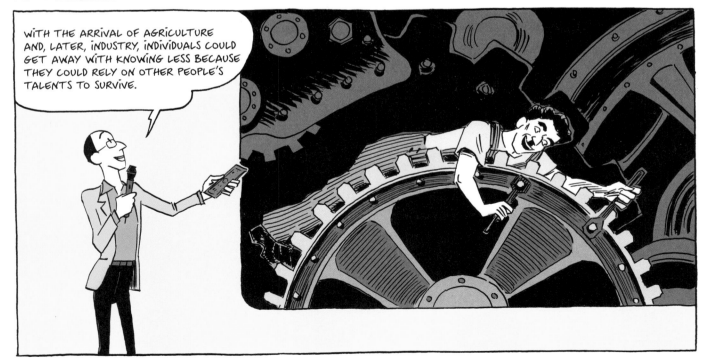

WITH THE ARRIVAL OF AGRICULTURE AND, LATER, INDUSTRY, INDIVIDUALS COULD GET AWAY WITH KNOWING LESS BECAUSE THEY COULD RELY ON OTHER PEOPLE'S TALENTS TO SURVIVE.

I'LL EXPLAIN WHAT I MEAN.

WE GENERALLY ASSUME THAT PEOPLE TODAY KNOW MUCH MORE THAN THEY DID IN ANCIENT TIMES...

OF COURSE, COLLECTIVELY OUR SOCIETIES KNOW A LOT MORE THAN ANCIENT HUNTER-GATHERERS.

BUT THE TRUTH IS EVERY INDIVIDUAL KNOWS LESS. I MEAN, WOULD YOU BE ABLE TO BUILD ALL THESE?

PEOPLE TODAY TEND TO KNOW HOW TO DO ONLY ONE SPECIFIC THING. SO, SOME OF THEM KNOW HOW TO WORK THE MACHINE THAT MAKES CAR TIRES. BUT THEY DON'T KNOW HOW TO MAKE THE ENGINE, THE STEERING WHEEL OR THE HEADLIGHTS.

BEING A HISTORIAN, I KNOW QUITE A LOT ABOUT HISTORY.

BUT I DON'T KNOW HOW TO HUNT OR HARVEST MY OWN FOOD, MAKE MY OWN CLOTHES OR BUILD MY OWN HOUSE...

IT'S THE SAME WITH EVERY PROFESSION. I HAVE ABSOLUTELY NO IDEA HOW TO FLY A PLANE!

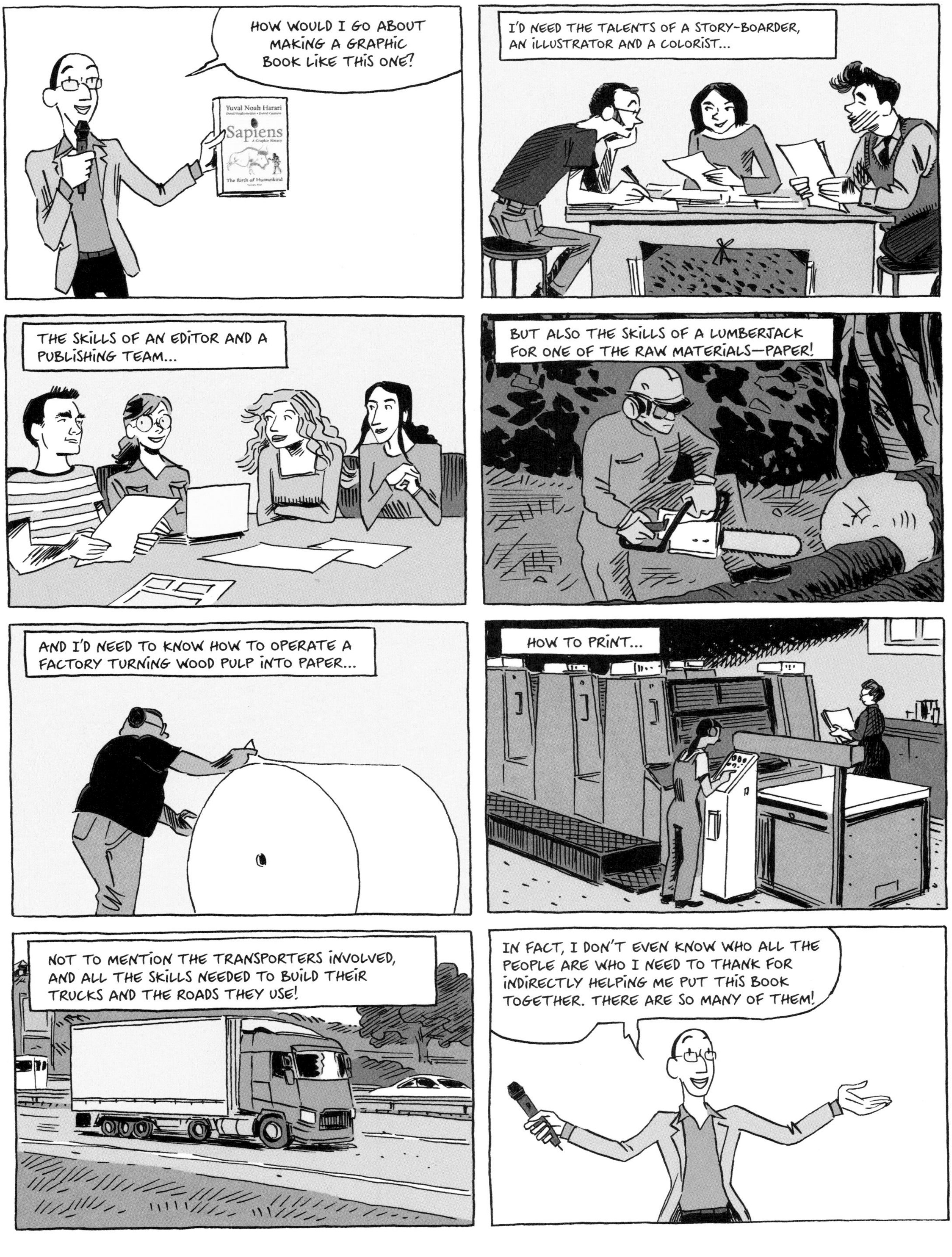
HOW WOULD I GO ABOUT MAKING A GRAPHIC BOOK LIKE THIS ONE?
Yuval Noah Harari
Sapiens
The Birth of Humankind
I'D NEED THE TALENTS OF A STORY-BOARDER, AN ILLUSTRATOR AND A COLORIST...
THE SKILLS OF AN EDITOR AND A PUBLISHING TEAM...
BUT ALSO THE SKILLS OF A LUMBERJACK FOR ONE OF THE RAW MATERIALS—PAPER!
AND I'D NEED TO KNOW HOW TO OPERATE A FACTORY TURNING WOOD PULP INTO PAPER...
HOW TO PRINT...
NOT TO MENTION THE TRANSPORTERS INVOLVED, AND ALL THE SKILLS NEEDED TO BUILD THEIR TRUCKS AND THE ROADS THEY USE!
IN FACT, I DON'T EVEN KNOW WHO ALL THE PEOPLE ARE WHO I NEED TO THANK FOR INDIRECTLY HELPING ME PUT THIS BOOK TOGETHER. THERE ARE SO MANY OF THEM!

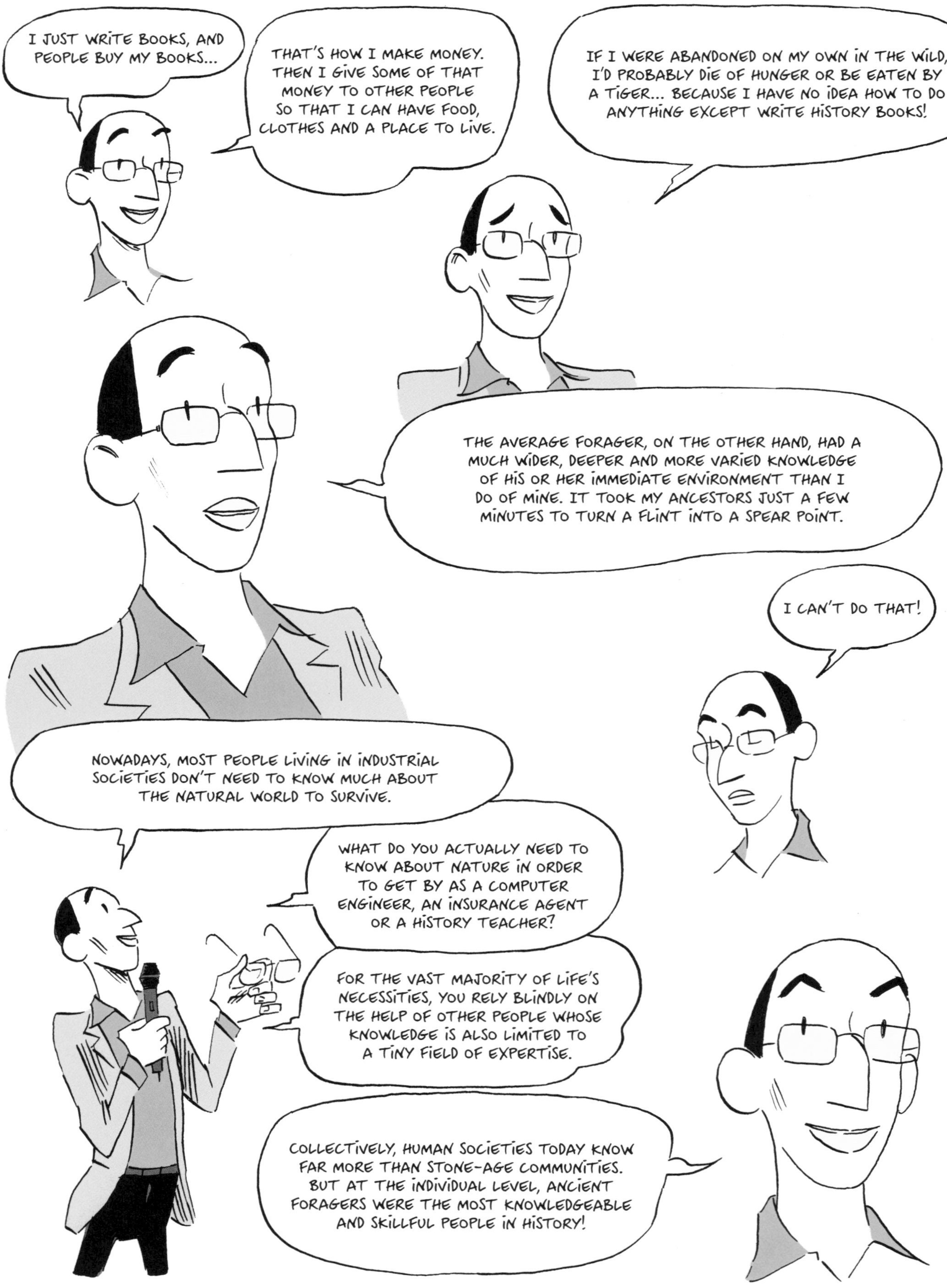
I JUST WRITE BOOKS, AND PEOPLE BUY MY BOOKS...
THAT'S HOW I MAKE MONEY. THEN I GIVE SOME OF THAT MONEY TO OTHER PEOPLE SO THAT I CAN HAVE FOOD, CLOTHES AND A PLACE TO LIVE.
IF I WERE ABANDONED ON MY OWN IN THE WILD, I'D PROBABLY DIE OF HUNGER OR BE EATEN BY A TIGER... BECAUSE I HAVE NO IDEA HOW TO DO ANYTHING EXCEPT WRITE HISTORY BOOKS!
THE AVERAGE FORAGER, ON THE OTHER HAND, HAD A MUCH WIDER, DEEPER AND MORE VARIED KNOWLEDGE OF HIS OR HER IMMEDIATE ENVIRONMENT THAN I DO OF MINE. IT TOOK MY ANCESTORS JUST A FEW MINUTES TO TURN A FLINT INTO A SPEAR POINT.
I CAN'T DO THAT!
NOWADAYS, MOST PEOPLE LIVING IN INDUSTRIAL SOCIETIES DON'T NEED TO KNOW MUCH ABOUT THE NATURAL WORLD TO SURVIVE.
WHAT DO YOU ACTUALLY NEED TO KNOW ABOUT NATURE IN ORDER TO GET BY AS A COMPUTER ENGINEER, AN INSURANCE AGENT OR A HISTORY TEACHER?
FOR THE VAST MAJORITY OF LIFE'S NECESSITIES, YOU RELY BLINDLY ON THE HELP OF OTHER PEOPLE WHOSE KNOWLEDGE IS ALSO LIMITED TO A TINY FIELD OF EXPERTISE.
COLLECTIVELY, HUMAN SOCIETIES TODAY KNOW FAR MORE THAN STONE-AGE COMMUNITIES. BUT AT THE INDIVIDUAL LEVEL, ANCIENT FORAGERS WERE THE MOST KNOWLEDGEABLE AND SKILLFUL PEOPLE IN HISTORY!

THANK YOU SO MUCH, YUVAL.

AND NOW, LADIES AND GENTLEMEN, I'D LIKE TO INVITE YOU TO JOIN US BACK HERE AFTER THE LUNCH BREAK.

CLAP CLAP CLAP CLAP CLAP

COME, PLEASE, FOLLOW ME. LET'S TRY THIS RESTAURANT THAT DR. DUARTE RECOMMENDED...

SO, ARE YOU SAYING THAT STONE-AGE HUNTER-GATHERERS HAD A MORE COMFORTABLE AND GRATIFYING WAY OF LIFE THAN THE POPULATIONS THAT CAME AFTER THEM?

WELL, IT DEPENDS WHO YOU COMPARE THEM WITH. MOST PEOPLE WHO "CAME AFTER THEM" WERE IMPOVERISHED PEASANTS RATHER THAN PEOPLE LIKE US FLYING TO CONFERENCES AND EATING IN SNAZZY RESTAURANTS!

IN TODAY'S WORLD MANY MILLIONS OF PEOPLE WORK UP TO 60 OR 80 HOURS A WEEK.

AND EVEN IN THE MOST DEVELOPED ECONOMIES, PEOPLE OFTEN WORK AN AVERAGE OF 40 TO 45 HOURS A WEEK.

BUT HUNTER-GATHERERS WHO CURRENTLY LIVE IN THE MOST INHOSPITABLE HABITATS, SUCH AS THE KALAHARI DESERT, WORK AN AVERAGE OF JUST 35 TO 45 HOURS A WEEK.

THEY HUNT JUST ONCE EVERY THREE DAYS, AND FORAGING TAKES, SAY, THREE TO SIX HOURS A DAY. IN NORMAL CONDITIONS, THAT'S ENOUGH TO FEED THE BAND.

WHICH MEANS THAT ANCIENT HUNTER-GATHERERS WHO LIVED IN MORE FERTILE AREAS POSSIBLY SPENT EVEN LESS TIME GATHERING FOOD AND RAW MATERIALS.

AND ANOTHER THING, THEY WOULDN'T HAVE HAD ANYTHING LIKE SO MANY HOUSEHOLD CHORES!

NO DISHES TO DO! AND NO VACUUMING, NO FLOORS TO POLISH OR DIAPERS TO CHANGE, AND DEFINITELY NO BILLS TO PAY!

THE FORAGER ECONOMY GAVE MOST PEOPLE MORE INTERESTING LIVES THAN AGRICULTURE OR INDUSTRY CAN OFFER!

IN OUR MODERN WORLD, A FACTORY WORKER CAN LEAVE HER HOUSE AT ABOUT SEVEN IN THE MORNING, TRUDGE THROUGH POLLUTED STREETS TO A WORKSHOP, AND THEN OPERATE THE SAME MACHINE ALL DAY LONG... TEN HOURS OF MIND-NUMBING WORK BEFORE RETURNING HOME AT ABOUT SEVEN IN THE EVENING TO FACE THE DISHES AND THE LAUNDRY!

30,000 YEARS AGO, A FORAGER MIGHT LEAVE CAMP WITH HER COMPANIONS AT MAYBE EIGHT IN THE MORNING. THEY'D ROAM THROUGH NEARBY FORESTS AND MEADOWS, PICKING MUSHROOMS, DIGGING FOR ROOT VEGETABLES, CATCHING FROGS...

BY EARLY AFTERNOON THEY'D BE BACK AT THE CAMP TO MAKE LUNCH!

Restaurante

THAT LEFT PLENTY OF TIME FOR PLAYING WITH THEIR KIDS OR JUST HANGING OUT!

OR FOR GOSSIPING AND TELLING STORIES!

THANK YOU.

BUT SURELY LIFE WASN'T SO IDYLLIC BACK THEN?

NATURALLY...

...SOMETIMES YOU COULD BE CAUGHT BY A BEAR OR BITTEN BY A SNAKE.

BUT AT LEAST THEY DIDN'T HAVE TO WORRY ABOUT TRAFFIC ACCIDENTS OR INDUSTRIAL POLLUTION!

IN MOST PLACES AND MOST OF THE TIME, FORAGING PROVIDED IDEAL NUTRITION.

AS A BIOLOGIST, I DON'T FIND THAT AT ALL SURPRISING. HUMANS LIVED AS FORAGERS FOR HUNDREDS OF THOUSANDS OF YEARS, SO THE HUMAN BODY WAS WELL ADAPTED TO A FORAGING DIET.

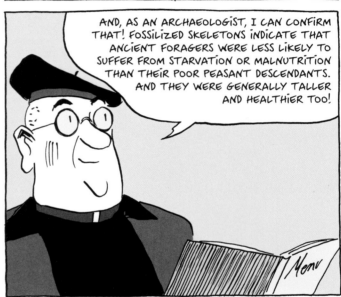

AND, AS AN ARCHAEOLOGIST, I CAN CONFIRM THAT! FOSSILIZED SKELETONS INDICATE THAT ANCIENT FORAGERS WERE LESS LIKELY TO SUFFER FROM STARVATION OR MALNUTRITION THAN THEIR POOR PEASANT DESCENDANTS. AND THEY WERE GENERALLY TALLER AND HEALTHIER TOO!

APPARENTLY, THEIR AVERAGE LIFE EXPECTANCY WAS ONLY ABOUT 30 OR 40 YEARS, BUT THAT'S MOSTLY DUE TO HIGH INFANT MORTALITY RATES. CHILDREN WHO MADE IT THROUGH THOSE PERILOUS EARLY YEARS HAD A GOOD CHANCE OF GETTING TO 50 OR 60—AS OLD AS 80 IN SOME CASES!

WE KNOW THAT IN MODERN FORAGER POPULATIONS, A WOMAN OF 45 CAN EXPECT TO LIVE ANOTHER 20 YEARS, AND AROUND 5% TO 8% OF THE POPULATION IS OVER 60.

THE SECRET OF THE FORAGERS' SUCCESS, THE THING THAT PROTECTED THEM FROM FAMINE AND MALNUTRITION, WAS THEIR VARIED DIET.

SUBSISTENCE FARMERS TEND TO HAVE A VERY LIMITED, UNBALANCED DIET. EVEN MORE SO IN PREMODERN TIMES WHEN AN AGRICULTURAL POPULATION WOULD GET MOST OF THEIR CALORIES FROM A SINGLE CROP...

WHEAT, POTATOES OR RICE... A DIET LIKE THAT DOESN'T HAVE ALL THE VITAMINS, MINERALS AND OTHER NUTRITIONAL ELEMENTS THAT HUMANS NEED.

TYPICAL PEASANTS IN TRADITIONAL CHINA ATE RICE FOR BREAKFAST, LUNCH AND DINNER. IF THEY WERE LUCKY, THEY MIGHT HAVE THE SAME THE FOLLOWING DAY.

AND, UNFORTUNATELY, THERE ARE STILL PEOPLE DYING OF MALNUTRITION IN SOME PARTS OF THE WORLD TODAY.

WHAT A CONTRAST TO ANCIENT FORAGERS WHO REGULARLY ATE DOZENS OF DIFFERENT FOODSTUFFS! THE ANCIENT ANCESTORS OF THOSE RICE-EATING PEASANTS MIGHT HAVE HAD BERRIES AND MUSHROOMS FOR BREAKFAST, THEN FRUITS, SNAILS AND TURTLES FOR LUNCH, AND RABBIT STEAK WITH WILD ONIONS FOR DINNER! THIS DIVERSITY MEANT THAT FORAGERS GOT ALL THE NUTRIENTS THEY NEEDED.

THERE WAS ANOTHER BENEFIT TOO: BECAUSE THEY WEREN'T DEPENDENT ON ANY ONE TYPE OF FOOD, THEY WEREN'T SO LIKELY TO GO HUNGRY IF A PARTICULAR FOOD SOURCE FAILED. AGRICULTURAL SOCIETIES ARE RAVAGED BY FAMINE IF DROUGHTS, FLOODS OR FIRES RUIN THEIR ANNUAL RICE OR POTATO CROP.

OH DEAR! I'M SORRY, I'M SUCH A KLUTZ!

BUT CAN WE ALL AGREE THAT FORAGER SOCIETIES WERE STILL THREATENED BY NATURAL DISASTERS? THEY MUST HAVE SUFFERED FAMINES TOO, DIDN'T THEY?

YES, OF COURSE, BUT THEY USUALLY FOUND IT EASIER TO COPE WITH THESE CALAMITIES. IF THEY LOST SOME OF THEIR STAPLE FOODS, THERE WERE OTHER SPECIES THEY COULD GATHER OR HUNT, OR THEY COULD MOVE TO A LESS AFFECTED AREA.

AND ANCIENT FORAGERS ALSO SUFFERED LESS FROM INFECTIOUS DISEASES! ISN'T THAT SO, PROFESSOR?

ABSOLU...

TCHOOO!

EXCUSE ME!

MOST OF OUR INFECTIOUS DISEASES—THINGS LIKE FLU, SMALLPOX AND MEASLES—CAME TO US FROM ANIMALS LIKE CHICKENS, COWS AND PIGS ONLY AFTER WE DOMESTICATED THEM IN THE AGRICULTURAL REVOLUTION.

OH, MY GOODNESS! WHAT ARE YOU DOING, KIKI?

AHA, YOU WANT SOMETHING TO EAT TOO... NO, KIKI, NOT IN A RESTAURANT.

SOME EPIDEMICS, LIKE COVID-19, COME FROM WILD ANIMALS, BUT ARE SPREAD BY MODERN TRANSPORT SYSTEMS.

ANCIENT FORAGERS DIDN'T FARM CHICKENS OR TRAVEL IN AIRPLANES, SO THEY LARGELY AVOIDED EPIDEMICS.

OH, ALRIGHT, JUST ONE LITTLE MOUTHFUL, THEN.

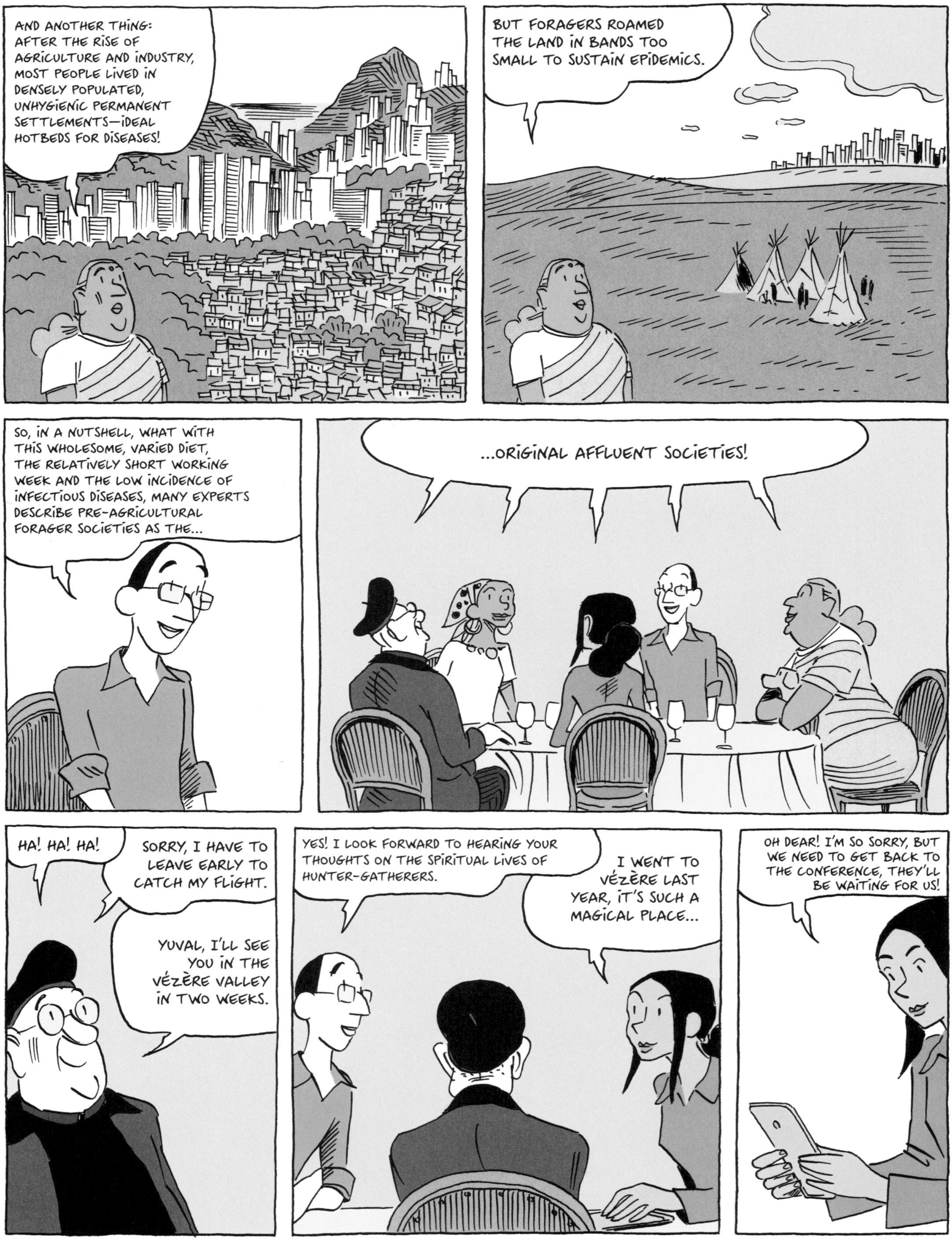
AND ANOTHER THING: AFTER THE RISE OF AGRICULTURE AND INDUSTRY, MOST PEOPLE LIVED IN DENSELY POPULATED, UNHYGIENIC PERMANENT SETTLEMENTS—IDEAL HOTBEDS FOR DISEASES!
BUT FORAGERS ROAMED THE LAND IN BANDS TOO SMALL TO SUSTAIN EPIDEMICS.
SO, IN A NUTSHELL, WHAT WITH THIS WHOLESOME, VARIED DIET, THE RELATIVELY SHORT WORKING WEEK AND THE LOW INCIDENCE OF INFECTIOUS DISEASES, MANY EXPERTS DESCRIBE PRE-AGRICULTURAL FORAGER SOCIETIES AS THE...
...ORIGINAL AFFLUENT SOCIETIES!
HA! HA! HA!
SORRY, I HAVE TO LEAVE EARLY TO CATCH MY FLIGHT.
YUVAL, I'LL SEE YOU IN THE VÉZÈRE VALLEY IN TWO WEEKS.
YES! I LOOK FORWARD TO HEARING YOUR THOUGHTS ON THE SPIRITUAL LIVES OF HUNTER-GATHERERS.
I WENT TO VÉZÈRE LAST YEAR, IT'S SUCH A MAGICAL PLACE...
OH DEAR! I'M SO SORRY, BUT WE NEED TO GET BACK TO THE CONFERENCE, THEY'LL BE WAITING FOR US!

IT WOULD BE MISLEADING TO IDEALIZE THE LIVES OF STONE-AGE HUNTER-GATHERERS.

THEY MAY WELL HAVE LIVED BETTER LIVES THAN MOST PEOPLE IN AGRICULTURAL AND INDUSTRIAL SOCIETIES, BUT THEIR WORLD COULD STILL BE HARSH AND UNFORGIVING.

SHORTAGES AND HARDSHIPS WEREN'T UNCOMMON, CHILD MORTALITY WAS HIGH AND AN ACCIDENT THAT WOULD BE MINOR TODAY COULD EASILY BECOME A DEATH SENTENCE.

MOST PEOPLE PROBABLY ENJOYED CLOSE LINKS WITHIN A ROAMING BAND, BUT ANYONE UNLUCKY ENOUGH TO INCUR HOSTILITY OR MOCKERY FROM THE BAND PROBABLY SUFFERED TERRIBLY.

MODERN FORAGERS HAVE BEEN KNOWN TO ABANDON AND EVEN KILL OLD OR DISABLED PEOPLE WHO CAN'T KEEP UP.

UNWANTED BABIES AND CHILDREN MAY ALSO BE KILLED, AND THERE ARE EVEN CASES OF RELIGIOUSLY INSPIRED HUMAN SACRIFICE.

I'D NOW LIKE TO INVITE YOU TO WATCH A DOCUMENTARY ON THIS VERY SUBJECT. I SHOULD WARN YOU, MANY OF YOU MAY FIND IT DISTURBING. DR. DUARTE WILL THEN HELP US TO CONTEXTUALIZE THE CONTENT, WHICH FOCUSES ON THE ACHÉ PEOPLE, A HUNTER-GATHERER SOCIETY WHOSE WAY OF LIFE ONLY RECENTLY DISAPPEARED. LIGHTS, PLEASE.

THE JUNGLES OF PARAGUAY. UP UNTIL THE 1960S, THIS WAS HOME TO A PEOPLE OF HUNTER-GATHERERS CALLED THE ACHÉ.

THEIR TRADITIONS AND THEIR WAY OF LIFE DISTURBED MANY ANTHROPOLOGISTS AT THE TIME. SOME OF THEIR CUSTOMS ARE VERY APPEALING, WHILE OTHERS ARE UNUSUALLY SHOCKING.

FOR EXAMPLE, WHEN A VALUED MEMBER OF A GROUP DIED, THE ACHÉ CUSTOM WAS TO KILL A LITTLE GIRL, AND BURY THE TWO TOGETHER.

ANTHROPOLOGISTS WHO INTERVIEWED THE ACHÉ IN THE 1960S RECORDED THE CASE OF A MIDDLE-AGED MAN WHO WAS ABANDONED BY HIS BAND WHEN HE FELL SICK AND COULD NOT KEEP UP WITH THE OTHERS.

THE ACHÉ LEFT HIM ALONE UNDER A TREE.

AT THE MERCY OF SCAVENGERS...

THE VULTURES WAITED A LONG TIME.

BUT THE MAN RECOVERED.

HE GOT TO HIS FEET AND WENT BACK TO HIS BAND.

ANOTHER MEMBER OF THE BAND FOUND A NEW NAME FOR HIM...

HEY, VULTURE POOP!

WHEN AN ACHÉ WOMAN BECAME OLD AND FRAIL, THE REST OF THE BAND SAW HER AS A BURDEN.

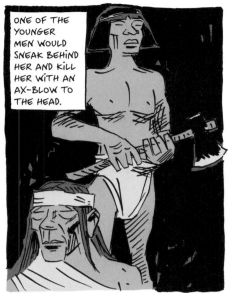

ONE OF THE YOUNGER MEN WOULD SNEAK BEHIND HER AND KILL HER WITH AN AX-BLOW TO THE HEAD.

ONE ACHÉ MAN ANSWERED MANY OF THE ANTHROPOLOGISTS' QUESTIONS WITH STORIES OF HIS PRIME YEARS IN THE JUNGLE.

I WAS THE ONE WHO KILLED OLD WOMEN. I KILLED MY AUNTS... WOMEN WERE FRIGHTENED OF ME...

NOW, HERE, WHERE THE WHITE PEOPLE LIVE, I'M A WEAKLING.

BABIES BORN WITH NO HAIR OR WHO WERE CONSIDERED UNDERDEVELOPED WERE KILLED IMMEDIATELY

MY FIRST BABY WAS KILLED BECAUSE THE MEN IN MY BAND DIDN'T WANT ANOTHER GIRL.

IN ANOTHER INCIDENT, A MAN KILLED A YOUNG BOY.

I WAS JUST IN A BAD MOOD AND HE WOULDN'T STOP CRYING.

ANOTHER CHILD WAS BURIED ALIVE. BECAUSE IT WAS "A FUNNY-LOOKING CHILD AND THE OTHER CHILDREN LAUGHED AT IT."

YOU MIGHT CONCLUDE THAT THE ACHÉ WERE BRUTAL CRIMINALS. BUT DON'T BE TOO QUICK TO JUDGE THEM. AS THIS ARCHIVE DOCUMENTARY SHOWS US, THE ACHÉ ALSO HAD POSITIVE QUALITIES. MOST OF THE TIME THEY WERE FRIENDLY TO ONE ANOTHER, THEY WERE ALWAYS SMILING AND LAUGHING.

ANTHROPOLOGISTS WHO LIVED WITH THEM FOR YEARS SAY THAT VIOLENCE BETWEEN ADULTS WAS VERY RARE. MEN AND WOMEN WERE FREE TO CHANGE PARTNERS AT WILL.

THEY HAD NO HIERARCHY OR LEADERS, AND GENERALLY AVOIDED DOMINEERING INDIVIDUALS. THEY WERE EXTREMELY GENEROUS WITH THEIR FEW POSSESSIONS, AND WEREN'T IN THE LEAST INTERESTED IN SUCCESS OR WEALTH.

THE THINGS THEY VALUED MOST IN LIFE WERE GOOD SOCIAL INTERACTIONS AND HIGH-QUALITY FRIENDSHIPS.

THEY VIEWED THE KILLING OF CHILDREN, THE SICK OR THE ELDERLY IN THE SAME WAY THAT MANY PEOPLE NOW VIEW ABORTION AND EUTHANASIA.

WE SHOULDN'T LOSE SIGHT OF THE FACT THAT THE ACHÉ WERE RELENTLESSLY HUNTED AND KILLED BY PARAGUAYAN FARMERS. THE DESPERATE NEED TO FLEE THEIR KILLERS COULD EXPLAIN THEIR EXCEPTIONALLY HARSH ATTITUDE TOWARD ANYONE WHO MIGHT BE A LIABILITY TO THE BAND.

THE TRUTH IS THAT, LIKE EVERY HUMAN SOCIETY, ACHÉ SOCIETY WAS VERY COMPLEX. WE SHOULDN'T RUSH TO DEMONIZE OR IDEALIZE THEM BASED ON SUCH LIMITED KNOWLEDGE OR INTERACTION. THEY WEREN'T ANGELS OR DEMONS—JUST HUMANS.

AND THAT'S EXACTLY WHAT STONE-AGE HUNTER-GATHERERS WERE TOO!

CLAP CLAP CLAP CLAP CLAP CLAP CLAP

THE VÉZÈRE VALLEY, FRANCE. TWO WEEKS LATER.

THIS MAGNIFICENT REGION IS HOME TO FASCINATING EVIDENCE OF THE SPIRITUAL AND EMOTIONAL LIFE OF THE ANCIENT HUNTER-GATHERERS. BUT UNFORTUNATELY, THIS EVIDENCE ISN'T EASY TO INTERPRET...

WE CAN FAIRLY CONFIDENTLY RECONSTRUCT THE BASICS OF THE ANCIENT FORAGERS' ECONOMY, USING VARIOUS OBJECTIVE, QUANTIFIABLE FACTORS.

FOR EXAMPLE, SCIENTISTS CAN CALCULATE HOW MANY CALORIES A PERSON NEEDED PER DAY TO SURVIVE...

HOW MANY CALORIES THEY COULD OBTAIN FROM A POUND OF WALNUTS...

AND HOW MANY WALNUTS THEY COULD HARVEST FROM A SQUARE MILE OF FOREST.

AND THIS DATA MEANS WE CAN MAKE AN EDUCATED GUESS ABOUT THE RELATIVE IMPORTANCE OF WALNUTS IN THEIR DIET.

BUT WHAT CAN WE SAY ABOUT THEIR SPIRITUAL AND MENTAL LIVES?

DID THEY CONSIDER WALNUTS A DELICACY, OR A HUMDRUM STAPLE?

DID THEY BELIEVE THAT WALNUT TREES WERE INHABITED BY SPIRITS?

DID THEY THINK THE LEAVES WERE PRETTY!

IF A FORAGER BOY WANTED TO TAKE A FORAGER GIRL TO A ROMANTIC SPOT, DID THE SHADE OF A WALNUT TREE DO THE TRICK?

IT'S FAR MORE DIFFICULT FOR US TO FIND OUT ABOUT THEIR THOUGHTS, BELIEFS AND FEELINGS—THAT'S INEVITABLE!

STILL, MOST SCHOLARS AGREE THAT WHAT WE CALL ANIMISTIC BELIEFS WERE COMMON AMONG ANCIENT FORAGERS.

HELLO-OOO! FATHER KLÜG! I'M HERE!

WAIT FOR ME, I'M COMING TO JOIN YOU!

ACTUALLY, FATHER KLÜG, COULD YOU EXPLAIN EXACTLY WHAT ANIMISM IS?

AHA, YES. YOU SEE THE WORD ANIMISM COMES FROM THE LATIN "ANIMA," WHICH MEANS SPIRIT OR SOUL...

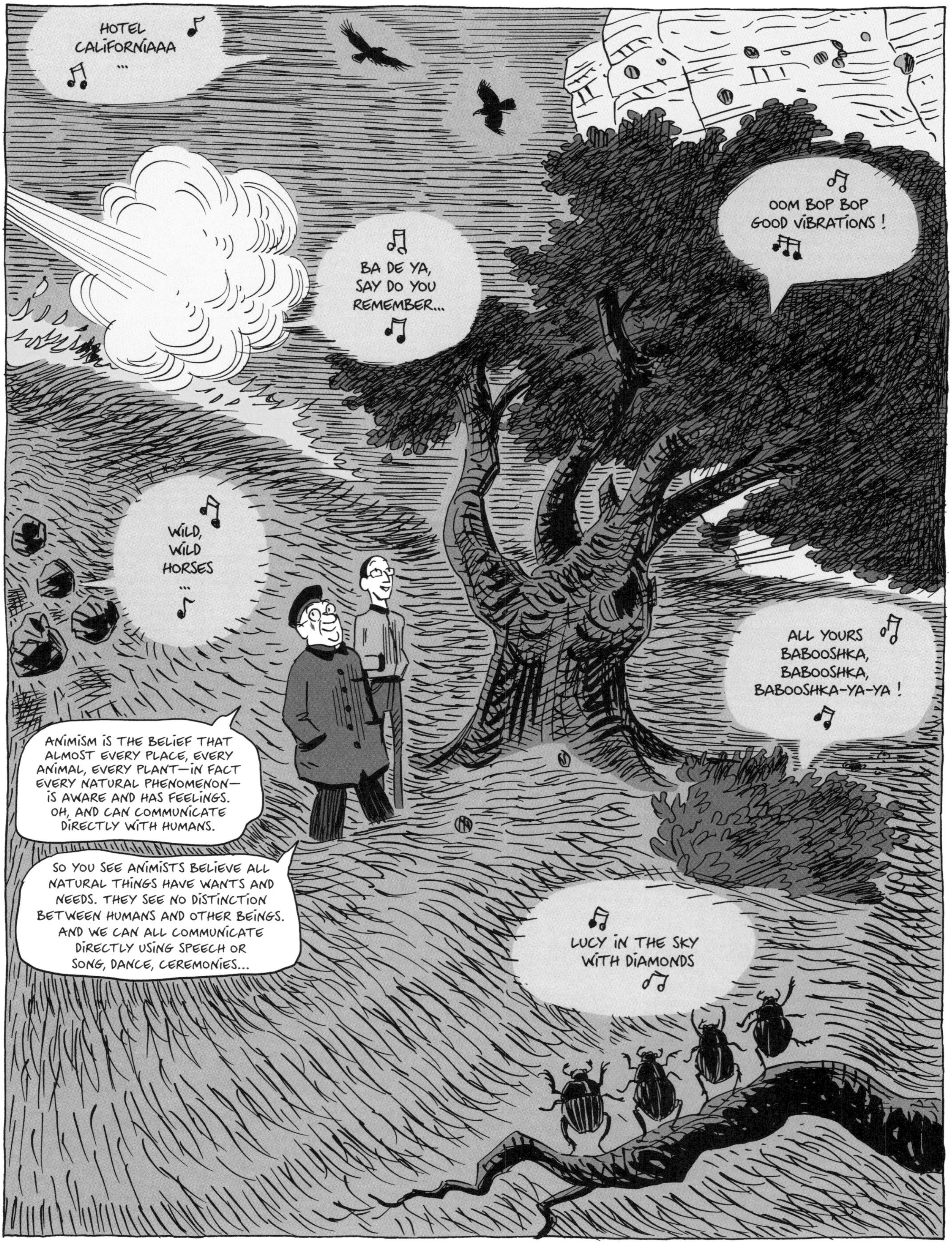
HOTEL CALIFORNIAAA ...
BA DE YA, SAY DO YOU REMEMBER...
OOM BOP BOP GOOD VIBRATIONS!
WILD, WILD HORSES ...
ALL YOURS BABOOSHKA, BABOOSHKA, BABOOSHKA-YA-YA!
ANIMISM IS THE BELIEF THAT ALMOST EVERY PLACE, EVERY ANIMAL, EVERY PLANT—IN FACT EVERY NATURAL PHENOMENON—IS AWARE AND HAS FEELINGS. OH, AND CAN COMMUNICATE DIRECTLY WITH HUMANS.
SO YOU SEE ANIMISTS BELIEVE ALL NATURAL THINGS HAVE WANTS AND NEEDS. THEY SEE NO DISTINCTION BETWEEN HUMANS AND OTHER BEINGS. AND WE CAN ALL COMMUNICATE DIRECTLY USING SPEECH OR SONG, DANCE, CEREMONIES...
LUCY IN THE SKY WITH DIAMONDS

THEY THOUGHT THAT A ROCK COULD BE ANGRY WITH PEOPLE ABOUT SOMETHING THEY DID AND HAPPY ABOUT SOMETHING ELSE.

IT COULD ADMONISH THEM OR ASK THEM A FAVOR. AND THE HUMANS THEMSELVES COULD TALK TO THE ROCK TO MOLLIFY IT... OR THREATEN IT.

IN THE ANIMIST WORLD, IT'S NOT JUST ROCKS, PLANTS AND ANIMALS THAT ARE ANIMATED.

THERE ARE ALSO INTANGIBLE ENTITIES LIKE THE SPIRITS OF THE DEAD, AND OTHER FRIENDLY— OR UNFRIENDLY!—SPIRITS, WHAT WE WOULD NOW CALL DEMONS, FAIRIES AND ANGELS.

A HUNTER MIGHT TALK TO A HERD OF DEER AND ASK FOR ONE TO SACRIFICE ITSELF. IF THE HUNT IS SUCCESSFUL, HE MIGHT ASK THE DEAD ANIMAL TO FORGIVE HIM.

WHEN SOMEONE FALLS SICK, A SHAMAN CAN CONTACT THE SPIRIT THAT CAUSED THE SICKNESS, AND TRY TO APPEASE IT. OR, IF THAT FAILS, TO FRIGHTEN IT AWAY!

IN A REAL TIGHT SPOT, THE SHAMAN COULD ASK FOR HELP FROM OTHER SPIRITS.

AND THERE'S SOMETHING ALL THESE COMMUNICATIONS HAVE IN COMMON:

THEY'RE NOT ADDRESSED TO UNIVERSAL GODS BUT TO LOCAL ENTITIES, PERHAPS A SPECIFIC TREE OR STREAM, OR A PARTICULAR GHOST.

SO, JUST AS THEY SEE NO DISTINCTION BETWEEN HUMANS AND OTHER BEINGS, THEY ALSO IMAGINE NO STRICT HIERARCHY.

NON-HUMAN ENTITIES AREN'T THERE JUST TO PROVIDE FOR HUMAN NEEDS.

AND THEY DON'T BELIEVE IN ALL-POWERFUL GODS, EITHER, THE SORT WHO RUN THE WORLD HOWEVER THEY PLEASE. THE ANIMIST WORLD DOESN'T REVOLVE AROUND HUMANS OR ANY OTHER PARTICULAR GROUP OF BEINGS.

ANIMISM ISN'T A SPECIFIC RELIGION. IT'S A GENERIC NAME FOR THOUSANDS OF DIFFERENT RELIGIONS, CULTS AND BELIEFS.

THE THING THAT MAKES THEM ALL "ANIMIST" IS THEIR APPROACH TO THE WORLD—AND TO MAN'S PLACE IN IT!

RIGHT, SO IS SAYING THAT ANCIENT FORAGERS WERE PROBABLY ANIMIST A BIT LIKE SAYING THAT PREMODERN FARMERS WERE MOSTLY THEISTS?

EXACTLY. AHA, YES, THE THEISTS—THAT WORD, I'M SURE YOU KNOW, COMES FROM THE GREEK WORD FOR GOD, "THEOS." NOW, THEISTS BELIEVE THAT THE WHOLE SET-UP HERE IS BASED ON A HIERARCHICAL RELATIONSHIP BETWEEN HUMANS AND A SMALL GROUP OF VERY POWERFUL ENTITIES: GODS.

IS IT SAFE TO SAY THAT PREMODERN FARMERS TENDED TO BE THEISTS?

YES, BUT THAT DOESN'T TELL US A GREAT DEAL... THE DEVIL'S IN THE DETAIL!

BECAUSE THE GENERIC TERM THEIST IS VERY BROAD. EACH OF THESE DIFFERENT GROUPS OF "THEISTS" THOUGHT THE OTHERS' BELIEFS AND PRACTICES WERE WEIRD AND HERETICAL. THE DIFFERENCES BETWEEN "ANIMIST" GROUPS OF FORAGERS WERE PROBABLY JUST AS BIG. THEIR RELIGIOUS EXPERIENCE MAY HAVE BEEN TURBULENT AND FULL OF CONTROVERSIES, REFORMS AND REVOLUTIONS.

Chinese bureaucrat
(1st century)

Roman legionary
(2nd century)

Viking warrior
(10th century)

Iranian Sufi
(12th century)

Aztec priest
(15th century)

Witch-burning Puritan,
Massachusetts
(17th century)

Jewish rabbi, Poland
(18th century)

STILL, THESE CAUTIOUS GENERALIZATIONS ARE ABOUT AS FAR AS WE CAN GO.

YOU SEE, WE'RE JUST VERY SHORT OF EVIDENCE. ANY ATTEMPT TO DESCRIBE THE SPECIFICS OF STONE-AGE SPIRITUALITY CAN'T HELP BEING HIGHLY SPECULATIVE.

WHAT LITTLE EVIDENCE WE DO HAVE—A HANDFUL OF ARTIFACTS AND A FEW CAVE PAINTINGS—ARE OPEN TO ALL SORTS OF INTERPRETATIONS. LIKE THESE... OH, MY GOODNESS, THE LASCAUX CAVE PAINTINGS... AREN'T THEY MAGNIFICENT!

HA, HA! SO WHEN SCHOLARS CLAIM THEY KNOW WHAT ANCIENT FORAGERS BELIEVED, IT OFTEN TELLS US MORE ABOUT THE SCHOLARS' OWN PRECONCEPTIONS THAN ABOUT STONE-AGE RELIGION!

QUITE RIGHT, YUVAL, QUITE RIGHT. WE SHOULDN'T BE MAKING MOUNTAINS OF THEORY OUT OF MOLEHILLS OF TOMB RELICS, CAVE PAINTINGS AND BONE STATUETTES! FAR BETTER TO BE HONEST AND ADMIT THAT WE ONLY HAVE THE HAZIEST OF NOTIONS ABOUT THE RELIGIONS OF OUR FORAGER ANCESTORS!

WE ASSUME THAT THEY WERE ANIMISTS, BUT WE DON'T KNOW WHAT SORT OF SPIRITS THEY PRAYED TO, WHAT FESTIVALS THEY CELEBRATED OR WHAT TABOOS THEY KEPT.

WE DON'T EVEN KNOW WHAT STORIES THEY TOLD, FOR GOODNESS' SAKE!

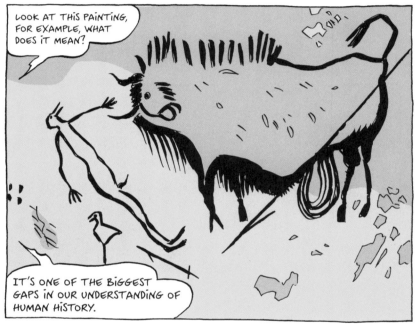

LOOK AT THIS PAINTING, FOR EXAMPLE, WHAT DOES IT MEAN?

IT'S ONE OF THE BIGGEST GAPS IN OUR UNDERSTANDING OF HUMAN HISTORY.

SOME PEOPLE SAY IT'S A MAN WITH A BIRD'S HEAD AND AN ERECT PENIS, AND HE'S BEING KILLED BY A BISON. THERE'S ANOTHER BIRD UNDER HIM, AND THAT MIGHT REPRESENT HIS SOUL BEING RELEASED FROM HIS BODY IN THE MOMENT OF DEATH.

IF THAT'S WHAT'S GOING ON, THIS PAINTING'S NOT ABOUT A HUNTING ACCIDENT BUT ABOUT THE SOUL PASSING FROM THIS WORLD TO THE NEXT... BUT WE HAVE ABSOLUTELY NO WAY OF KNOWING IF THIS SPECULATION IS VALID!

THERE ARE MANY OTHER THEORIES ABOUT THIS PAINTING, ALL OF THEM EQUALLY PLAUSIBLE.

HAH! IT'S LIKE AN INKBLOT TEST— IT TELLS US MORE ABOUT MODERN RESEARCHERS' MINDSETS THAN THE ORIGINAL PAINTERS' RELIGIOUS BELIEFS!

THE SAME GOES FOR THE FORAGERS' SOCIO-POLITICAL WORLD. WE KNOW HARDLY ANYTHING ABOUT THAT EITHER.

SPECIALISTS CAN'T EVEN AGREE ON THE BASICS! I MEAN, DID THEY HAVE PRIVATE PROPERTY? NUCLEAR FAMILIES? MONOGAMOUS RELATIONSHIPS?!

DIFFERENT BANDS PROBABLY HAD DIFFERENT STRUCTURES.

I AGREE, SOME COULD HAVE BEEN AS HIERARCHICAL AND VIOLENT AS THE NASTIEST CHIMPANZEE GROUP, AND OTHERS AS LAID-BACK, PEACEFUL AND, ERM, LASCIVIOUS AS A BUNCH OF BONOBOS.

IN SUNGIR IN RUSSIA, ARCHAEOLOGISTS FOUND A 34,000-YEAR-OLD BURIAL SITE FROM A MAMMOTH-HUNTING CULTURE.

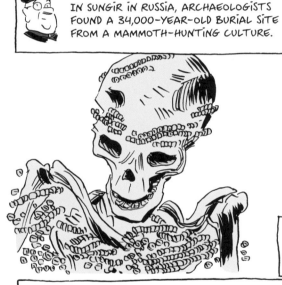

ONE GRAVE HAD THE SKELETON OF A 40-YEAR-OLD MAN COVERED WITH STRINGS OF BEADS MADE FROM MAMMOTH IVORY.

THAT GRAVE CONTAINED ABOUT 3,000 BEADS IN ALL. THE DEAD MAN WAS WEARING A HAT DECORATED WITH FOX TEETH, AND HE HAD 25 IVORY BRACELETS ON HIS WRISTS.

OTHER GRAVES AT THE SITE HAD FEWER ACCESSORIES... SO SCHOLARS DEDUCED THAT THESE SUNGIR MAMMOTH-HUNTERS LIVED IN A HIERARCHICAL SOCIETY, AND THAT THE DEAD MAN WAS HEAD OF A BAND, PERHAPS EVEN OF A TRIBE MADE UP OF SEVERAL BANDS.

IS IT POSSIBLE THAT JUST A FEW DOZEN MEMBERS OF ONE BAND COULD HAVE MADE ALL THOSE ACCESSORIES ???

WELL, IT SEEMS UNLIKELY.

ANYWAY, THE ARCHAEOLOGISTS THEN FOUND AN EVEN MORE INTERESTING GRAVE, WHERE TWO YOUNG BOYS HAD BEEN BURIED HEAD TO HEAD. ONE WAS 12 OR 13, THE OTHER 9 OR 10.

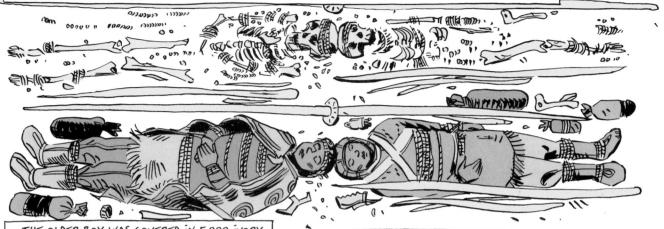

THE OLDER BOY WAS COVERED IN 5,000 IVORY BEADS! HE WAS ALSO WEARING A FOX-TOOTH HAT AND A BELT WITH 250 FOX TEETH ON IT—I TELL YOU, IT WOULD HAVE TAKEN AT LEAST 60 DEAD FOXES TO PRODUCE THAT MANY!

THE YOUNGER BOY WAS COVERED WITH 5,250 IVORY BEADS, AND BOTH SKELETONS WERE SURROUNDED BY STATUETTES AND IVORY OBJECTS.

NOW, IT WOULD HAVE TAKEN A SKILLED ARTISAN ABOUT 45 MINUTES TO MAKE A SINGLE IVORY BEAD! IN OTHER WORDS, THE 10,000 BEADS ON THOSE CHILDREN WOULD HAVE TAKEN 7,500 HOURS OF DELICATE WORK—NOT TO MENTION ALL THE OTHER OBJECTS???!

BUT THAT'S WELL OVER THREE YEARS' WORK!

THAT'S INCREDIBLE! WAIT, SURELY, IT'S VERY UNLIKELY THAT AT SUCH A YOUNG AGE, THESE SUNGIR BOYS COULD HAVE BEEN LEADERS OR MAMMOTH-HUNTERS...

EXACTLY... THE ONLY THING THAT CAN EXPLAIN WHY A COUPLE OF CHILDREN WERE GIVEN SUCH AN EXTRAVAGANT BURIAL IIIIIS... CULTURAL BELIEFS.

THERE ARE THREE DIFFERENT THEORIES ABOUT THE SUNGIR BOYS... THE FIRST IS THAT THEY OWED THEIR ELEVATED STANDING TO THEIR PARENTS.

THEY COULD HAVE BEEN THE TRIBE LEADER'S CHILDREN... BUT RECENT DNA ANALYSIS HAS NEGATED THAT: IT SHOWED THAT THEY WEREN'T BROTHERS AND THE 40-YEAR-OLD MAN WASN'T THEIR FATHER.

THE SECOND THEORY SUGGESTS THAT WHEN THEY WERE BORN THEIR COMMUNITY IDENTIFIED THEM AS INCARNATIONS OF LONG-DEAD SPIRITS.

BUT THEN ALONG COMES THE THIRD THEORY, WHICH ARGUES THAT THEIR LAVISH BURIAL HAD NOTHING TO DO WITH THEIR STATUS... IT WAS ALL ABOUT HOW THEY DIED.

THEY COULD HAVE BEEN RITUALLY SACRIFICED, PERHAPS AS PART OF THE LEADER'S BURIAL RIGHTS, AND THEN GIVEN A BIG, FANCY BURIAL.

WHATEVER ACTUALLY HAPPENED, THOSE GRAVES IN SUNGIR ARE SOME OF THE BEST EVIDENCE WE HAVE ABOUT HOW SAPIENS LIVED 34,000 YEARS AGO: THEY ALREADY CONGREGATED IN LARGE TRIBES AND THEY'D COME UP WITH SOCIOPOLITICAL CODES THAT WENT WAY BEYOND THE DICTATES OF OUR DNA... AND BEYOND BEHAVIORS SEEN IN OTHER HUMAN OR ANIMAL SPECIES.

AND NOW WE COME TO THE THORNY SUBJECT OF VIOLENCE AND WAR.

SOME SPECIALISTS ARGUE THAT ANCIENT FORAGERS WERE VERY PEACEFUL...

...AND THAT WAR AND VIOLENCE BEGAN ONLY AFTER THE AGRICULTURAL REVOLUTION.

OTHERS SAY THAT THE ANCIENT FORAGERS' WORLD WAS HORRIBLY CRUEL AND VIOLENT.

BOTH SCHOOLS OF THOUGHT ARE LIKE CASTLES IN THE AIR, CONNECTED TO REALITY BY A FEW MEAGER ARCHAEOLOGICAL REMAINS AND SOME QUESTIONABLE ANTHROPOLOGICAL OBSERVATIONS.

AS PROFESSOR YOSHITA EXPLAINED SO WELL IN RIO, THE ANTHROPOLOGICAL EVIDENCE IS FASCINATING BUT, TO SAY THE LEAST, PROBLEMATIC.

AND ANYWAY, IN RECENT YEARS FORAGERS HAVE BEEN INCREASINGLY POLICED BY THE AUTHORITY OF MODERN STATES THAT PREVENT ANY LARGE-SCALE CONFLICTS.

MODERN DAY FORAGERS MOSTLY LIVE IN ISOLATED, INHOSPITABLE AREAS LIKE THE ARCTIC OR THE KALAHARI, PLACES WITH VERY LOW POPULATION DENSITY AND LIMITED OPPORTUNITIES FOR FIGHTING OTHER PEOPLE.

MODERN DAY SCHOLARS HAVE HAD ONLY TWO OPPORTUNITIES TO OBSERVE LARGE, RELATIVELY DENSE POPULATIONS OF INDEPENDENT FORAGERS...

IN NORTHWESTERN NORTH AMERICA IN THE 19TH CENTURY...

AND IN NORTHERN AUSTRALIA IN THE 19TH AND EARLY 20TH CENTURY.

IN BOTH INSTANCES, THERE WERE FREQUENT ARMED CONFLICTS... BUT HOW MUCH OF THE BIGGER PICTURE CAN WE DEDUCE FROM JUST TWO CASES?

BESIDES, BOTH THE AMERINDIANS AND THE ABORIGINAL AUSTRALIANS WERE INFLUENCED AND PRESSURED BY EUROPEAN INVADERS. SO, YOU NEVER KNOW, MAYBE THEIR VIOLENCE WAS A RESULT OF EUROPEAN IMPERIALISM.

AS FOR CONCRETE PROOF OF WAR DURING THE STONE AGE... WELL, WE HAVE PRECIOUS LITTLE OF IT.

BUT ABSENCE OF EVIDENCE IS NOT EVIDENCE OF ABSENCE! WHAT CLUES COULD BE LEFT BY A WAR THAT MIGHT HAVE BEEN FOUGHT 30,000 YEARS AGO?

OH DEAR, NONE AT ALL! THEY HAD NO FORTIFICATIONS OR WALLS BACK THEN, NO SWORDS OR EVEN SHIELDS.

WE DO SOMETIMES FIND SPEARHEADS FROM THAT ERA...

THIS ONE MAY WELL HAVE BEEN USED IN WAR...

...BUT IT COULD JUST AS EASILY HAVE BEEN USED FOR HUNTING.

YES, I TAKE YOUR POINT! AND I WOULD HAVE THOUGHT THAT FOSSILIZED HUMAN BONES ARE JUST AS DIFFICULT TO INTERPRET...

INDEED THEY ARE! LET'S IMAGINE THAT WE HAPPEN TO FIND A FOSSILIZED FRACTURED SKULL.

THE FRACTURE COULD BE FROM A WAR WOUND...

...OR FROM, I DON'T KNOW, A HUNTING ACCIDENT.

BUT THEN THE ABSENCE OF FRACTURES AND CUTS ON A FOSSILIZED BONE DOESN'T PROVE FOR SURE THAT THE PERSON DIED A PEACEFUL DEATH.

I MEAN, TRAUMA TO SOFT TISSUES CAN BE FATAL, AND IT LEAVES NO TRACE ON BONE.

THERE'S SOMETHING ELSE, THOUGH! IN PRE-INDUSTRIAL WARFARE, MORE THAN 90% OF DEATHS WERE CAUSED BY STARVATION, COLD AND DISEASE, NOT BY WEAPONS.

IMAGINE A SCENE 30,000 YEARS AGO, WITH ONE TRIBE DEFEATING THEIR NEIGHBORS AND DRIVING THEM OUT OF VALUABLE FORAGING GROUNDS.

LET'S SAY THAT TEN MEMBERS OF THE DEFEATED TRIBE ARE KILLED IN THE DECISIVE BATTLE.

THE NEXT YEAR ANOTHER HUNDRED MEMBERS OF THE DEFEATED TRIBE THEN DIE OF HUNGER, COLD OR DISEASE.

EVEN IF BY SOME MIRACLE ARCHAEOLOGISTS FIND ALL 110 SKELETONS, THEY MIGHT WRONGLY CONCLUDE THAT ONLY 10% OF THOSE PEOPLE WERE KILLED BY WAR.

HOW COULD WE KNOW THAT ALL OF THEM WERE ACTUALLY VICTIMS OF A MERCILESS CONFLICT?

RIGHT, BUT HAVING CONSIDERED ALL THE PROBLEMS, WHAT CAN YOU TELL US FOR SURE ABOUT THE ARCHAEOLOGICAL EVIDENCE THAT WE HAVE?

WHEN SCIENTISTS STUDIED 400 SKELETONS FROM THE PERIOD JUST BEFORE THE AGRICULTURAL REVOLUTION—THIS WAS IN PORTUGAL, I SEEM TO THINK—ONLY 2 SKELETONS SHOWED CLEAR MARKS OF VIOLENCE.

HMM... SUBSTANTIAL LESION HERE. WERE YOU IN A FIGHT?

NO, DOC. I, UM, I SLIPPED.

A SIMILAR STUDY OF MORE THAN 400 SKELETONS FROM THE SAME PERIOD—IN ISRAEL THIS TIME—CAME UP WITH JUST ONE CRACK IN ONE SKULL THAT COULD BE ATTRIBUTED TO HUMAN VIOLENCE.

WELL, DOESN'T ALL THIS SUPPORT THE THEORY THAT ANCIENT FORAGERS WERE PEACEFUL?

YES, BUT THERE ARE OTHER SURVEYS. AT THE INDIAN KNOLL SITE IN KENTUCKY, 48 OF THE 880 ANCIENT HUNTER-GATHERERS' SKELETONS SHOWED EVIDENCE OF VIOLENCE, LIKE MUTILATIONS AND EMBEDDED ARROWHEADS.

SOMEONE STEPS ONTO THE ROADWAY, OBSTRUCTING MY ROUTE.

A: I OFFER TO HELP HIM CARRY HIS GOODS HOME.

B: I PULL OVER TO LET HIM PASS.

C: I BLOW HIS BRAINS OUT.

48 OUT OF 880 MAY NOT SOUND LIKE A LOT... BUT IT'S ACTUALLY A VERY HIGH PERCENTAGE.

IF ALL 48 OF THOSE PEOPLE REALLY DID DIE VIOLENT DEATHS, THAT MEANS THAT 5.5% OF DEATHS IN THE AREA WERE CAUSED BY HUMAN VIOLENCE!

THE GLOBAL AVERAGE TODAY IS JUST 1.5%, AND THAT'S LUMPING WARS AND CRIMES TOGETHER!

IN THE 20TH CENTURY—AND LET'S BE HONEST, IT SAW THE BLOODIEST WARS AND MOST MASSIVE GENOCIDES IN HISTORY—ONLY 5% OF HUMAN DEATHS RESULTED FROM HUMAN VIOLENCE.

SO ANCIENT KENTUCKY MAY WELL HAVE BEEN AS VIOLENT AS THE 20TH CENTURY!

AT JEBEL SAHABA IN SUDAN, A 12,000-YEAR-OLD CEMETERY WITH 59 SKELETONS WAS FOUND.

24 OF THEM—THAT'S A HUGE 40%—SHOW SIGNS OF VIOLENCE, LIKE SKULL FRACTURES OR BONES WITH EMBEDDED STONE SPEARHEADS.

THE MOST ANCIENT SITE IN EUROPE WITH CLEAR EVIDENCE OF WAR IS A CEMETERY NEAR THE RIVER DNIEPER, IN UKRAINE, THAT DATES FROM 12,000 YEARS AGO. 5 OF THE 19 PEOPLE BURIED THERE SEEM TO HAVE COME TO A VIOLENT END.

SIMILAR SITES HAVE BEEN FOUND ALL OVER THE WORLD, FROM SCANDINAVIA TO CALIFORNIA, BUT NONE OF THEM GO BACK MORE THAN 12,000 YEARS.

AND WHERE DOES THAT LEAVE US? WHAT BEST REPRESENTS THE ANCIENT FORAGERS' WORLD— THE UNSCATHED SKELETONS IN PORTUGAL OR THE ABATTOIR IN JEBEL SAHABA?

HMM, THE ANSWER IS NEITHER. THERE'S JUST NO FIXED RATE FOR HUMAN VIOLENCE.

AS WE'VE SEEN, FORAGERS HAD QUITE AN ARRAY OF RELIGIONS AND SOCIAL STRUCTURES... WELL, THEY PROBABLY HAD VERY DIFFERENT RATES OF VIOLENCE TOO.

SOME PLACES AND TIMES WERE AS PEACEFUL AS PRESENT-DAY SWEDEN, OTHERS WERE AS VIOLENT AS THE SYRIA OF TODAY.

THANK YOU FOR EXPLAINING ALL THAT SO CLEARLY, FATHER KLÜG.

LET'S HAVE A RECAP! WE REALLY DON'T KNOW VERY MUCH ABOUT THE LIVES OF ANCIENT FORAGERS...

WE CAN CERTAINLY MAKE A FEW GENERAL DEDUCTIONS. BUT SPECIFIC EVENTS ARE STILL TANTALIZINGLY OUT OF REACH...

WHAT HAPPENED WHEN A SAPIENS BAND FIRST ENTERED A VALLEY INHABITED BY NEANDERTHALS? THE FOLLOWING YEARS MIGHT HAVE WITNESSED BREATHTAKING HISTORICAL DRAMAS.

UNFORTUNATELY, NOTHING WOULD HAVE SURVIVED FROM SUCH AN ENCOUNTER EXCEPT, IN THE BEST CASE SCENARIO, A FEW FOSSILIZED BONES AND A HANDFUL OF STONE TOOLS... AND THEY DON'T DIVULGE ALL THEIR SECRETS EVEN UNDER THE MOST INTENSE SCHOLARLY INQUISITIONS.

THEY MAY GIVE US INFORMATION ABOUT ANATOMY AND GENETICS...

AH! THIS PERSON WAS A REDHEAD!

HOW THINGS WERE DONE...

DIET...

JUDGING BY LASER ANALYSIS OF HIS TARTAR, THIS GUY ATE A LOT OF GAME FOR MANY YEARS, BUT THEN LIVED OFF NOTHING BUT DRIED FRUIT FOR THE LAST FEW MONTHS OF HIS LIFE...

EVEN SOCIAL STRUCTURES...

BUT NONE OF THAT TELLS US ANYTHING ABOUT THE POLITICAL ALLIANCES FORGED BETWEEN NEIGHBORING SAPIENS BANDS, OR THE SPIRITS OF THE DEAD THAT BLESSED THIS ALLIANCE, OR THE IVORY BEADS SECRETLY GIVEN TO THE LOCAL WITCH-DOCTOR TO SECURE THAT SPIRITUAL BLESSING.

THIS CURTAIN OF SILENCE HANGS OVER MOST OF HUMAN HISTORY.

TENS OF THOUSANDS OF YEARS THAT MAY HAVE SEEN WARS AND REVOLUTIONS, ECSTATIC RELIGIOUS MOVEMENTS, PROFOUND PHILOSOPHICAL THEORIES AND INCREDIBLE ARTISTIC MASTERPIECES.

FORAGERS MAY HAVE HAD ALL-CONQUERING NAPOLEONS RULING EMPIRES HALF THE SIZE OF LUXEMBOURG...

GIFTED BEETHOVENS WHO DIDN'T HAVE SYMPHONY ORCHESTRAS BUT MADE PEOPLE WEEP WITH THE TUNES THEY PLAYED ON BAMBOO FLUTES...

OR CHARISMATIC PROPHETS WHO REVEALED THE PRONOUNCEMENTS OF A LOCAL OAK TREE RATHER THAN A UNIVERSAL CREATOR GOD.

BUT I'M JUST GUESSING HERE! THE CURTAIN OF SILENCE IS SO THICK THAT WE CAN'T BE SURE THAT ANY OF THIS HAPPENED.

THAT'S RIGHT...

RESEARCHERS DO HAVE THIS HABIT OF ONLY ASKING QUESTIONS THAT THEY CAN REASONABLY EXPECT TO ANSWER!

188

WE'LL PROBABLY NEVER KNOW WHAT THE ANCIENT FORAGERS BELIEVED OR WHAT POLITICAL UPHEAVALS THEY LIVED THROUGH.

UNLESS, OF COURSE, SCIENTISTS MANAGE TO MAKE A TIME MACHINE...

OR SPIRITUALISTS ARRANGE SEANCES WITH OUR DISTANT ANCESTORS...

BUT IT'S STILL VITAL TO ASK QUESTIONS THAT HAVE NO ANSWERS. OTHERWISE, WE MIGHT BE TEMPTED TO DISMISS MOST OF HUMAN HISTORY WITH A GLIB "NAH, THE PEOPLE WHO LIVED BACK THEN DIDN'T DO ANYTHING IMPORTANT!"

THE TRUTH IS THEY DID ALL SORTS OF IMPORTANT THINGS. THEY SHAPED THE WORLD AROUND US... FAR MORE THAN MOST OF US REALIZE.

TREKKERS VISITING THE SIBERIAN TUNDRA...

THE DESERTS OF CENTRAL AUSTRALIA...

...AND THE AMAZONIAN RAINFOREST...

...THINK THEY'RE SEEING PRISTINE LANDSCAPES, VIRTUALLY UNTOUCHED BY HUMAN HAND. BUT THAT'S AN ILLUSION.

THE FORAGERS WERE THERE LONG BEFORE US, AND THEY BROUGHT ABOUT DRAMATIC CHANGES EVEN IN THE DENSEST JUNGLES AND THE MOST DESOLATE WILDERNESSES.

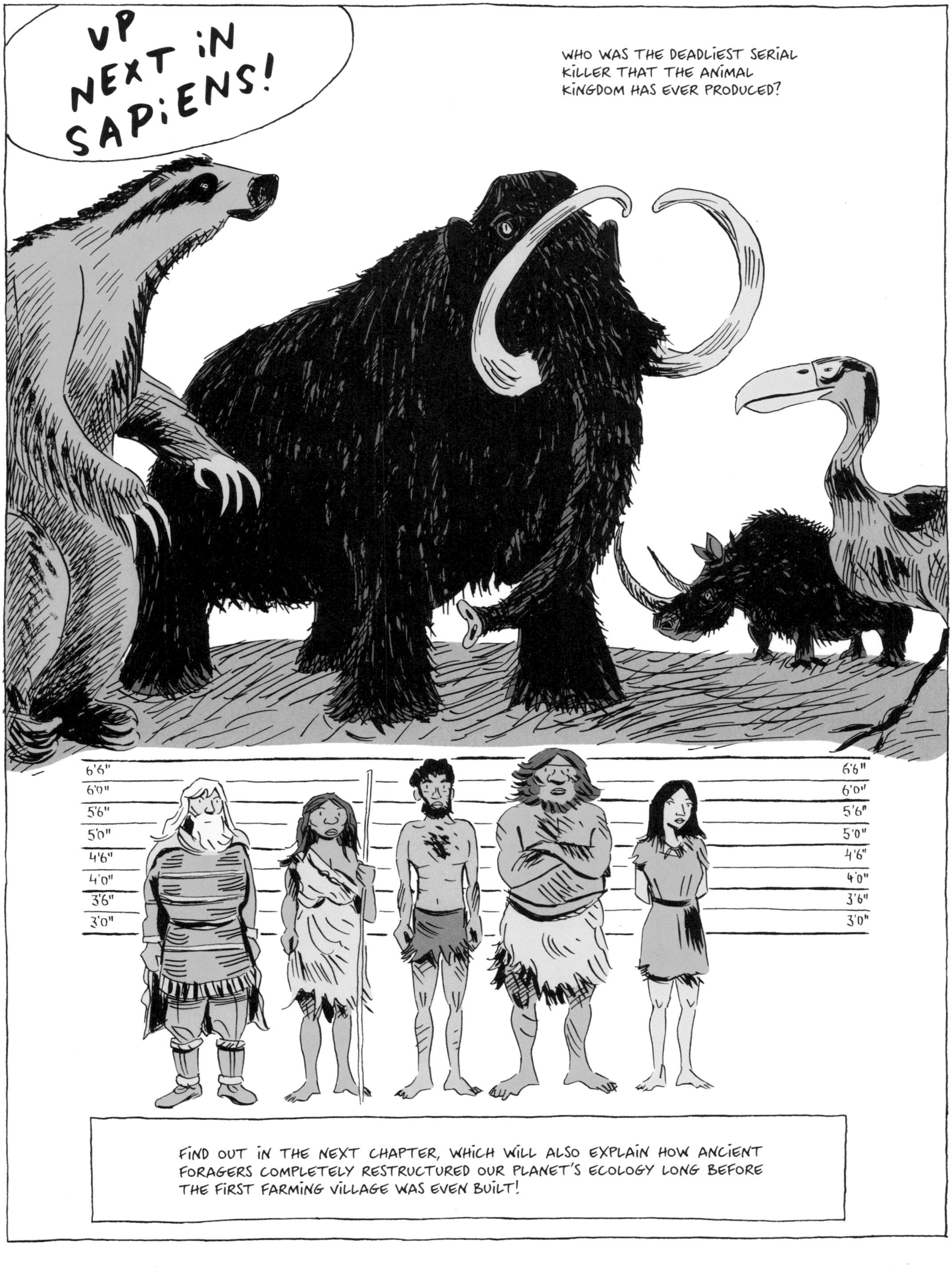
UP NEXT IN SAPIENS!
WHO WAS THE DEADLIEST SERIAL KILLER THAT THE ANIMAL KINGDOM HAS EVER PRODUCED?
6'6"
6'0"
5'6"
5'0"
4'6"
4'0"
3'6"
3'0"
FIND OUT IN THE NEXT CHAPTER, WHICH WILL ALSO EXPLAIN HOW ANCIENT FORAGERS COMPLETELY RESTRUCTURED OUR PLANET'S ECOLOGY LONG BEFORE THE FIRST FARMING VILLAGE WAS EVEN BUILT!

INTERCONTINENTAL
SERIAL KILLERS

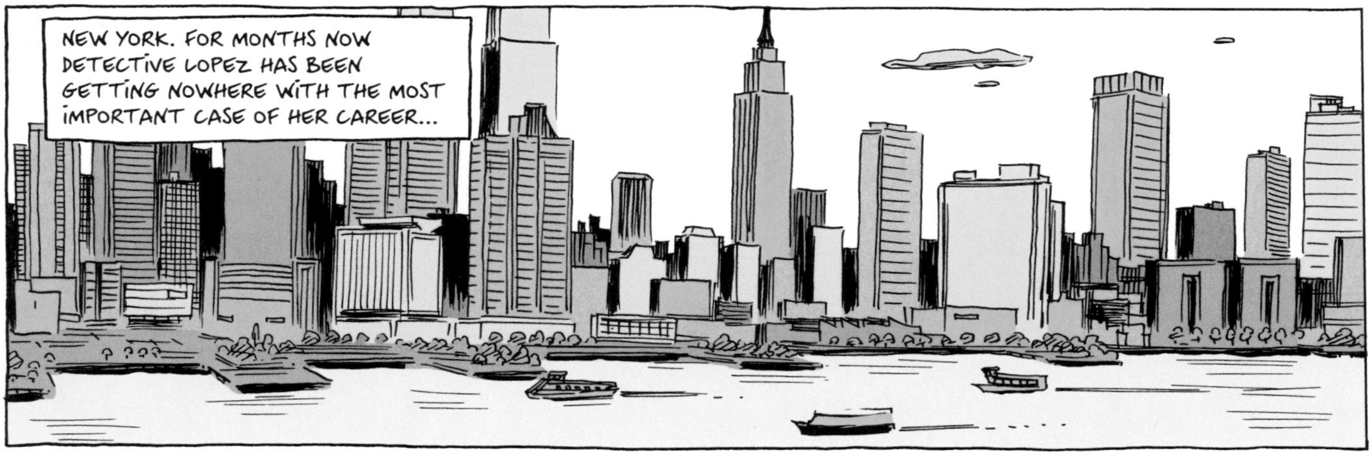
NEW YORK. FOR MONTHS NOW DETECTIVE LOPEZ HAS BEEN GETTING NOWHERE WITH THE MOST IMPORTANT CASE OF HER CAREER...

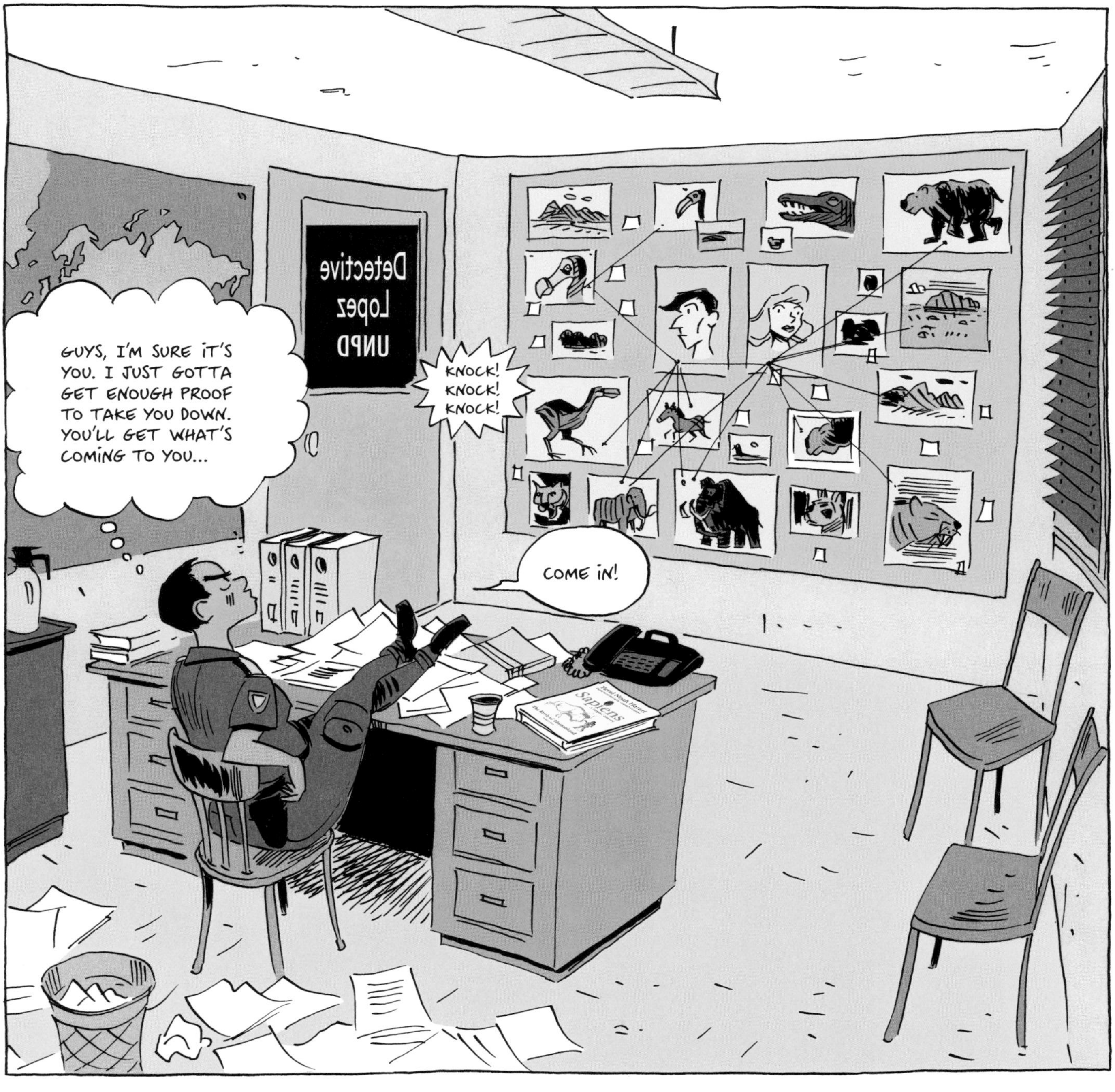
Detective Lopez UNPD
GUYS, I'M SURE IT'S YOU. I JUST GOTTA GET ENOUGH PROOF TO TAKE YOU DOWN. YOU'LL GET WHAT'S COMING TO YOU...
KNOCK! KNOCK! KNOCK!
COME IN!
Sapiens

THE PROFILERS ARE HERE, DETECTIVE LOPEZ.

SHOW 'EM IN!

COME ON IN.

MORNING! GREAT TIMING, I WAS HOPING TO SEE YOU!

THANK YOU, BUT I THINK THERE'S BEEN A MISTAKE... PROFESSOR HARARI AND I AREN'T PROFILERS AT ALL AND I'M AFRAID...

OK, OK, I KNOW, I WAS EXPECTING YOU.

THANKS FOR COMING, PROFESSOR.

PROFILER'S JUST THE WORD IN OUR LINGO. WHAT I REALLY NEED IS YOUR EXPERTISE TO KICK MY INQUIRY INTO SHAPE...

YOU WANT TEA? COFFEE!

AH! THERE'S MY LICORICE STICK! HELL, I MISS MY CIGARETTES.

UHM...TEA, THANKS.

194

I'M ON THE CASE OF THE CENTURY... WAIT WHAT? OF THE MILLENNIUM. BUT I JUST NEED SOME MORE PROOF TO THROW THE WORST SERIAL KILLERS OF ALL TIME IN JAIL.

IS THAT ALL? AND YOU'RE SURE WE'RE...

SURE I'M SURE. YOU'LL SEE.

YOU SEE THESE TWO, RIGHT? THEY'RE MY SUSPECTS. I HAVE THEM DETAINED HERE IN THE PRECINCT.

I 100% BELIEVE THEY'RE THE WORLD'S WORST ECOLOGICAL SERIAL KILLERS.

DAMAGE TO THE GLOBAL ECOSYSTEM, MASS MURDER OF HUNDREDS OF HUMAN AND ANIMAL SPECIES... OBVIOUSLY, THEY DENY EVERYTHING AND THEY WON'T TALK. BUT WHEREVER THESE GUYS GO, A WHOLE BUNCH OF BODIES ALWAYS SHOW UP.

WE'RE TALKING ABOUT A SERIOUSLY BAD COUPLE. THEY SAY THEY'RE CALLED BILL AND CINDY SAPIENS. LET'S CALL 'EM SAPIENS, OK?

NEXT TO THEM BONNIE & CLYDE LOOK LIKE A COUPLE OF CHOIRBOYS!

YES, I'VE HEARD OF THEM. BUT UM... DO YOU REALLY THINK THESE TWO PEOPLE COULD HAVE DONE EVERYTHING YOU SAY ON THEIR OWN?

OBVIOUSLY NOT. THEY'RE PART OF A HUGE CRIMINAL RING OF ORGANIZED GANGS. BUT THEY'RE THE ONLY TWO I GOT IN THE SLAMMER. IF I CAN GET THEM TO SING, THEY'LL LEAD ME TO THE OTHERS.

FIRST, LET ME TELL YOU HOW I FOUND THEM. IT WAS ACTUALLY THANKS TO YOUR GRAPHIC HISTORY THAT I SUDDENLY SAW THE LIGHT.

STOP ME IF I GET THIS STUFF WRONG...

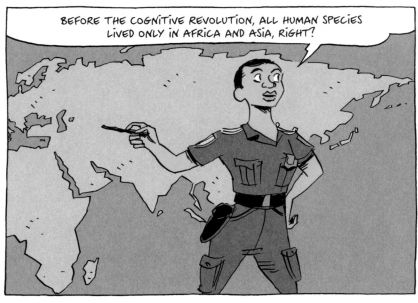

BEFORE THE COGNITIVE REVOLUTION, ALL HUMAN SPECIES LIVED ONLY IN AFRICA AND ASIA, RIGHT!

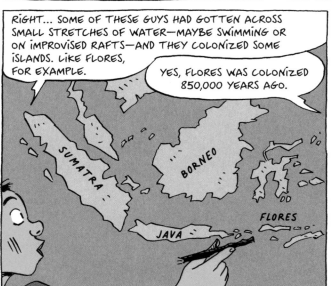

RIGHT... SOME OF THESE GUYS HAD GOTTEN ACROSS SMALL STRETCHES OF WATER—MAYBE SWIMMING OR ON IMPROVISED RAFTS—AND THEY COLONIZED SOME ISLANDS. LIKE FLORES, FOR EXAMPLE.

YES, FLORES WAS COLONIZED 850,000 YEARS AGO.

BUT THESE WERE KIND OF SPECIAL CASES. BACK THEN, HUMANS COULDN'T REALLY HEAD OUT TO SEA. NONE OF THESE SUCKERS MADE IT TO AMERICA, AUSTRALIA OR REMOTE ISLANDS LIKE, SAY, MADAGASCAR, NEW ZEALAND OR HAWAII. WE STILL ALL GOOD HERE!

THAT'S EXACTLY RIGHT. THE SEA ACTED AS A BARRIER TO HUMANS BUT ALSO TO MANY OTHER AFRO-ASIAN ANIMALS AND PLANTS, STOPPING THEM REACHING THE "OUTER WORLD."

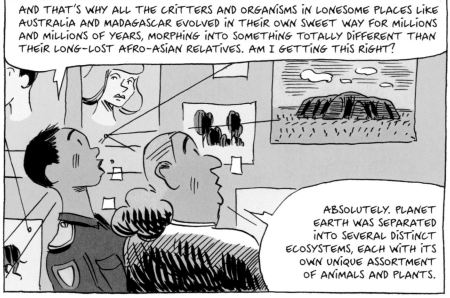

AND THAT'S WHY ALL THE CRITTERS AND ORGANISMS IN LONESOME PLACES LIKE AUSTRALIA AND MADAGASCAR EVOLVED IN THEIR OWN SWEET WAY FOR MILLIONS AND MILLIONS OF YEARS, MORPHING INTO SOMETHING TOTALLY DIFFERENT THAN THEIR LONG-LOST AFRO-ASIAN RELATIVES. AM I GETTING THIS RIGHT?

ABSOLUTELY. PLANET EARTH WAS SEPARATED INTO SEVERAL DISTINCT ECOSYSTEMS, EACH WITH ITS OWN UNIQUE ASSORTMENT OF ANIMALS AND PLANTS.

OK! WELL, LISTEN UP NOW: MY THEORY IS THAT THESE NASTY SAPIENS AND THEIR GANG CALLED TIME ON THIS... "BIOLOGICAL EXUBERANCE" I THINK YOU CALLED IT!

OK, HERE'S THE FIRST CRIME SCENE.

AUSTRALIA

TOTAL BUMMER, THIS WAS 50,000 YEARS AGO. SO THERE'S NOT A WHOLE HEAP OF EVIDENCE LEFT.

PROBLEM NUMBER ONE, IT'S HARD TO EXPLAIN HOW THESE SAPIENS GANGS EVEN MADE IT TO AUSTRALIA, OR WHERE THEY CAME FROM...

AUSTRALIA

I'M GUESSING THAT 50,000 YEARS AGO THEY ALREADY HAD FISHING VILLAGES ON THE INDONESIAN COAST AND PEOPLE HAD GOTTEN AROUND TO BUILDING RAFTS OR BOATS. BUT THAT'S JUST SPECULATION. SO FAR NOBODY FOUND ANY VILLAGES OR BOATS.

NOT EVEN AN OAR!

THAT'S HARDLY SURPRISING...

BOATS AND OARS ARE MADE OF WOOD AND BAMBOO, MATERIALS THAT CAN'T LAST TENS OF THOUSANDS OF YEARS IN THE TROPICS. AND ANYWAY, RISING SEA LEVELS HAVE BURIED THE ANCIENT INDONESIAN SHORELINE UNDER 330 FEET OF OCEAN.

SO WE HAVE NO DIRECT PROOF THAT SAPIENS COULD BUILD RAFTS AND BOATS AT THE TIME, OR NAVIGATE ON THE OPEN SEA.

BUT THERE IS CIRCUMSTANTIAL EVIDENCE THAT MIGHT HELP OUR INVESTIGATION, ESPECIALLY THE FACT THAT NOT LONG AFTER AUSTRALIA WAS SETTLED, OTHER SAPIENS MANAGED TO COLONIZE A LOT OF SMALL ISOLATED ISLANDS TO THE NORTH OF IT.

AUSTRALIA

SOME OF THEM, LIKE BUKA AND MANUS, WERE 125 MILES FROM THE CLOSEST LAND. IT'S HARD TO BELIEVE THAT ANYONE COULD HAVE REACHED AND COLONIZED MANUS WITHOUT SOPHISTICATED VESSELS AND SAILING SKILLS...

WE ALSO HAVE FIRM EVIDENCE OF REGULAR SEA TRADE BETWEEN SOME OF THESE ISLANDS, LIKE NEW IRELAND AND NEW BRITAIN...

NEW IRELAND

NEW BRITAIN

PAPUA

AUSTRALIA

THE UNITED STATES OF AMERICA
ONE DOLLAR

EXCELLENT! WE'RE MAKING PROGRESS! AND THERE'S PROOF OF THIS STUFF?

YES, AFTER THE COGNITIVE REVOLUTION SAPIENS DEVELOPED THE TECHNOLOGY, THE ORGANIZATIONAL SKILLS, AND PERHAPS EVEN THE VISION NECESSARY TO BREAK OUT OF AFRO-ASIA AND COLONIZE THE OUTER WORLD.

THIS WAS A REMARKABLE ACHIEVEMENT. TO REACH AUSTRALIA, THESE HUMANS HAD TO CROSS SEVERAL SEA CHANNELS, SOME MORE THAN 60 MILES WIDE. AND WHEN THEY LANDED THEY HAD TO ADAPT VIRTUALLY OVERNIGHT TO A COMPLETELY NEW ECOSYSTEM.

THE MOST REASONABLE THEORY GOES LIKE THIS: THE INDONESIAN ISLANDS ARE SEPARATED FROM ASIA AND FROM EACH OTHER BY NARROW STRAITS SO THE SAPIENS WHO LIVED THERE ABOUT 50,000 YEARS AGO PROBABLY DEVELOPED THE FIRST SEAFARING SOCIETIES. THEY LEARNED TO BUILD AND MANEUVER OCEAN-GOING VESSELS AND BECAME LONG-DISTANCE FISHERMEN, TRADERS AND EXPLORERS.

THIS WOULD HAVE BROUGHT ABOUT AN UNPRECEDENTED TRANSFORMATION IN HUMAN CAPABILITIES AND LIFESTYLES. LOTS OF NEW OPPORTUNITIES OPENED UP TO THEM...

YEAH RIGHT! LOTS OF OPPORTUNITIES TO TURN INTO TOTAL SERIAL KILLERS...

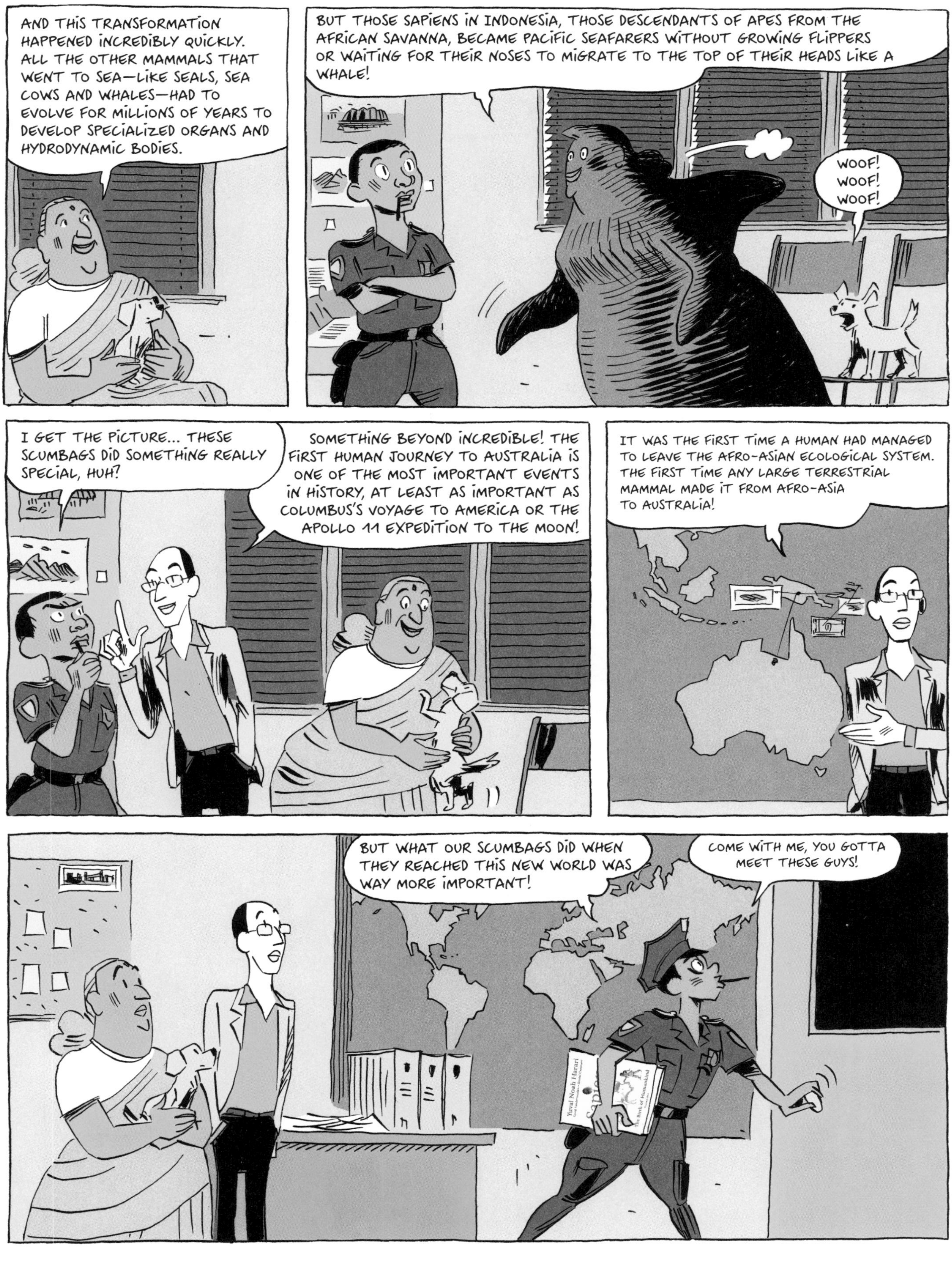
AND THIS TRANSFORMATION HAPPENED INCREDIBLY QUICKLY. ALL THE OTHER MAMMALS THAT WENT TO SEA—LIKE SEALS, SEA COWS AND WHALES—HAD TO EVOLVE FOR MILLIONS OF YEARS TO DEVELOP SPECIALIZED ORGANS AND HYDRODYNAMIC BODIES.
BUT THOSE SAPIENS IN INDONESIA, THOSE DESCENDANTS OF APES FROM THE AFRICAN SAVANNA, BECAME PACIFIC SEAFARERS WITHOUT GROWING FLIPPERS OR WAITING FOR THEIR NOSES TO MIGRATE TO THE TOP OF THEIR HEADS LIKE A WHALE!
WOOF! WOOF! WOOF!
I GET THE PICTURE... THESE SCUMBAGS DID SOMETHING REALLY SPECIAL, HUH?
SOMETHING BEYOND INCREDIBLE! THE FIRST HUMAN JOURNEY TO AUSTRALIA IS ONE OF THE MOST IMPORTANT EVENTS IN HISTORY, AT LEAST AS IMPORTANT AS COLUMBUS'S VOYAGE TO AMERICA OR THE APOLLO 11 EXPEDITION TO THE MOON!
IT WAS THE FIRST TIME A HUMAN HAD MANAGED TO LEAVE THE AFRO-ASIAN ECOLOGICAL SYSTEM. THE FIRST TIME ANY LARGE TERRESTRIAL MAMMAL MADE IT FROM AFRO-ASIA TO AUSTRALIA!
BUT WHAT OUR SCUMBAGS DID WHEN THEY REACHED THIS NEW WORLD WAS WAY MORE IMPORTANT!
COME WITH ME, YOU GOTTA MEET THESE GUYS!
Yuval Noah Harari
Sapiens
The Birth of Humankind

THE MINUTE THEY SET FOOT ON AN AUSTRALIAN BEACH, THESE HUSTLERS CLIMBED RIGHT TO THE TOP OF THE FOOD CHAIN. ALPHA DOGS! NUMERO UNO! CAPO DI TUTTI CAPI! PLANET EARTH WAS IN SHOCK.

YES, UP UNTIL THEN HUMANS HAD SHOWN SOME PRETTY INNOVATIVE ADAPTATIONS AND BEHAVIORS, BUT THEY'D HAD A NEGLIGIBLE EFFECT ON THEIR ENVIRONMENT. THEY'D BEEN REMARKABLY SUCCESSFUL AT MOVING INTO AND ADJUSTING TO NEW HABITATS... BUT WITHOUT DRASTICALLY CHANGING THEM.

HERE, TAKE A PEEK, THEY'RE IN THERE.

THESE TWO DIDN'T JUST MOVE TO AUSTRALIA AND ADAPT TO IT... OH NO! THEY COMPLETELY TRANSFORMED THE ECOSYSTEM.

SO, I'M GUESSING YOU STILL DON'T WANNA COOPERATE...

STILL NOT TALKING? WHAT ELSE IS NEW?

LOOK GUYS, THINGS AIN'T LOOKING TOO PRETTY FOR YOU.

PROFESSOR SARASWATI, DID YOU BRING THAT FILM I ASKED YOU FOR?

YES, OF COURSE, IT'S IN MY BAG. WHERE DID I PUT THAT MEMORY STICK?

OK, THANKS, PROF!

I CAN'T FIND ANY PROOF THAT YOU REACHED AUSTRALIA BY BOAT... YOUR FOOTPRINTS IN THE SAND WERE ERASED RIGHT AWAY BY THE WAVES.

BUT MOVING INLAND, YOU LEFT TOTALLY DIFFERENT FOOTPRINTS THAT HAVE NEVER BEEN ERASED!

THAT'S RIGHT! YOU DIDN'T COVER ALL YOUR BASES!

HERE'S WHAT HAPPENED, LET ME REFRESH YOUR MEMORY...

AS YOU TRAVELED FURTHER, YOU FOUND THIS MONDO WEIRDO WORLD FULL OF FREAKY ANIMALS! LIKE THIS MAGNIFICENT SPECIMEN, A 6-FOOT-5, 440-POUND KANGAROO. RINGING ANY BELLS?

AND THIS MARSUPIAL LION THE SIZE OF A MODERN TIGER. THIS GUY WAS THE BIGGEST AUSSIE PREDATOR BACK THEN!

PLUS THERE WERE KOALAS WAY TOO BIG TO BE CUDDLY AND CUTE...

AND FLIGHTLESS BIRDS TWICE THE SIZE OF AN OSTRICH THAT COULD SPRINT LIKE USAIN BOLT ON A TRIPLE ESPRESSO!

DRAGON-LIZARD THINGS AND 16-FOOT SNAKES HUNG OUT IN THE UNDERGROWTH.

THE GIANT DIPROTODON, A TWO-AND-A-HALF-TON WOMBAT, CRUISED AROUND IN THE FOREST.

THESE BEAUTIFUL GIANTS WERE BASICALLY UNKNOWN IN AFRICA AND ASIA, BUT BOY, WERE THEY THE KINGS IN AUSTRALIA!

APART FROM THE BIRDS AND REPTILES, ALL THESE CRITTERS WERE MARSUPIALS: LIKE KANGAROOS, THEY GAVE BIRTH TO HELPLESS, ITTY-BITTY BABIES, AND THEN RAISED THEM ON MILK IN A POUCH.

WHAT HAPPENED TO THEM, HUH, HUH?!?

ALMOST ALL THESE GIANTS VANISHED IN A FEW THOUSAND YEARS! TWENTY-THREE OF THE TWENTY-FOUR AUSTRALIAN ANIMALS WEIGHING MORE THAN 100 POUNDS... BOOM! GOODBYE FOREVER!

AND A LOT OF SMALLER SPECIES BOUGHT IT TOO! IT WAS CARNAGE, MESSING WITH THE WHOLE AUSSIE ECOSYSTEM! ALL THE FOOD CHAINS WERE BROKEN AND REARRANGED! THE ECOSYSTEM HADN'T SEEN ANYTHING LIKE IT FOR MILLIONS OF YEARS.

WHO'S TO BLAME IF NOT YOU TWO? COME ON, 'FESS UP!

WHAT THE HELL IS GOING ON?!?

SO NOW YOU'RE INTERROGATING MY CLIENTS WITHOUT ME?

OH GREAT!... NOT ADAMSKI...

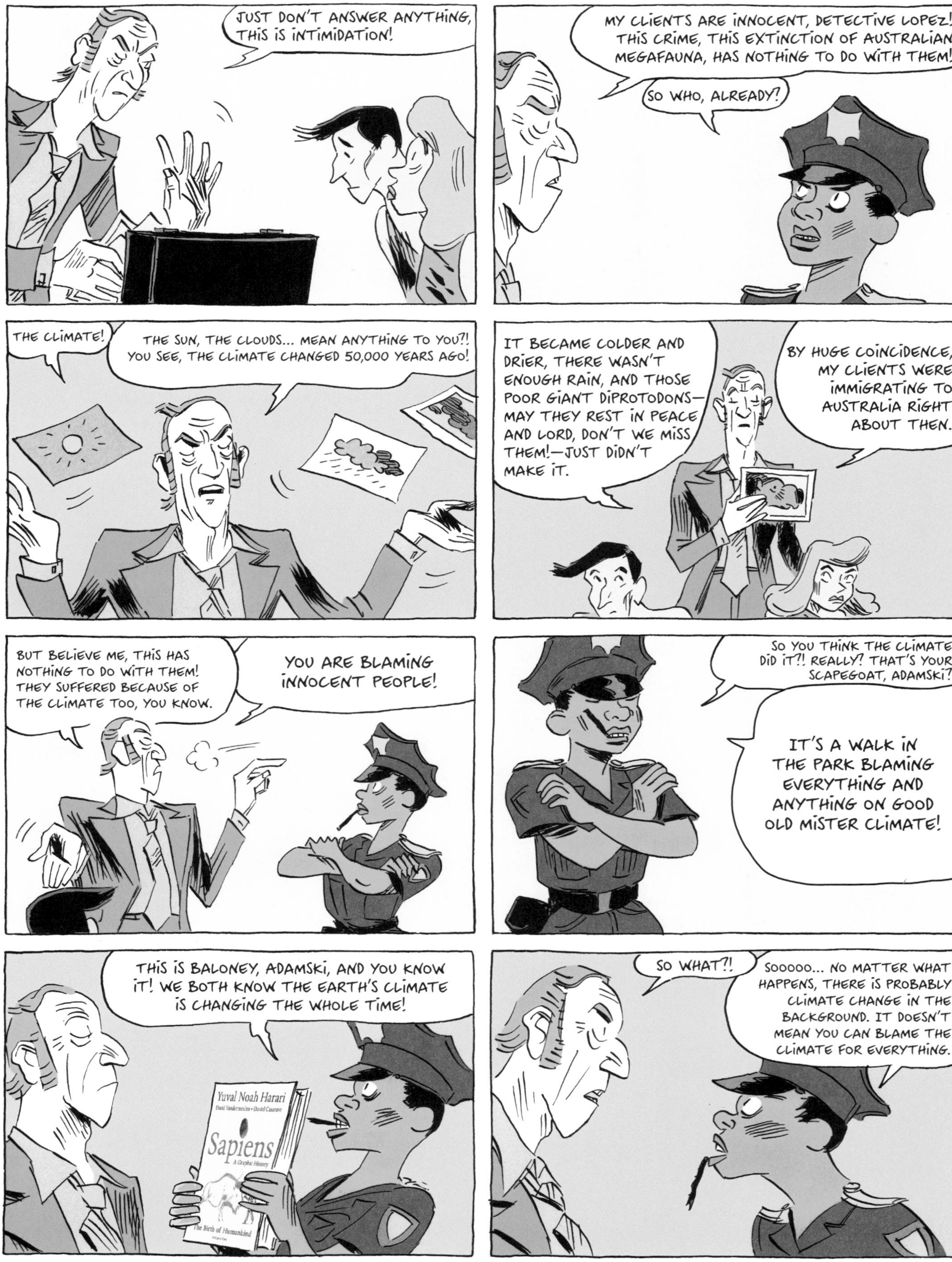
JUST DON'T ANSWER ANYTHING, THIS IS INTIMIDATION!
MY CLIENTS ARE INNOCENT, DETECTIVE LOPEZ! THIS CRIME, THIS EXTINCTION OF AUSTRALIAN MEGAFAUNA, HAS NOTHING TO DO WITH THEM!
SO WHO, ALREADY?
THE CLIMATE!
THE SUN, THE CLOUDS... MEAN ANYTHING TO YOU?! YOU SEE, THE CLIMATE CHANGED 50,000 YEARS AGO!
IT BECAME COLDER AND DRIER, THERE WASN'T ENOUGH RAIN, AND THOSE POOR GIANT DIPROTODONS—MAY THEY REST IN PEACE AND LORD, DON'T WE MISS THEM!—JUST DIDN'T MAKE IT.
BY HUGE COINCIDENCE, MY CLIENTS WERE IMMIGRATING TO AUSTRALIA RIGHT ABOUT THEN.
BUT BELIEVE ME, THIS HAS NOTHING TO DO WITH THEM! THEY SUFFERED BECAUSE OF THE CLIMATE TOO, YOU KNOW.
YOU ARE BLAMING INNOCENT PEOPLE!
SO YOU THINK THE CLIMATE DID IT?! REALLY? THAT'S YOUR SCAPEGOAT, ADAMSKI?
IT'S A WALK IN THE PARK BLAMING EVERYTHING AND ANYTHING ON GOOD OLD MISTER CLIMATE!
THIS IS BALONEY, ADAMSKI, AND YOU KNOW IT! WE BOTH KNOW THE EARTH'S CLIMATE IS CHANGING THE WHOLE TIME!
Yuval Noah Harari
Sapiens
A Graphic History
The Birth of Humankind
SO WHAT?!
SOOOOO... NO MATTER WHAT HAPPENS, THERE IS PROBABLY CLIMATE CHANGE IN THE BACKGROUND. IT DOESN'T MEAN YOU CAN BLAME THE CLIMATE FOR EVERYTHING.

FIRST OFF, EVEN IF YOU'RE RIGHT ABOUT AUSTRALIA'S CLIMATE CHANGING 50,000 YEARS AGO, IT WASN'T EXACTLY ARMAGEDDON! I CAN'T SEE HOW A LITTLE CHANGE IN THE WEATHER COULD HAVE CAUSED THIS MOTHER OF AN EXTINCTION!

IF YOU SAY SO! WERE YOU THERE?

I KNOW A WHOLE LOT MORE ABOUT THIS THAN YOU DO, ADAMSKI!

CLICK!

THIS PLANET HAS QUITE A HABIT OF COOLING OFF AND HEATING UP. IN THE LAST MILLION YEARS THERE'S BEEN AN ICE AGE ABOUT EVERY 100,000 YEARS. THE LATEST ONE WAS BETWEEN 75,000 AND 15,000 YEARS AGO AND IT WENT REAL CRAZY A COUPL'A TIMES: FIRST ABOUT 70,000 YEARS AGO AND THEN ABOUT 20,000 YEARS AGO.

YOUR LITTLE CHANGES 50,000 YEARS AGO WERE LIKE MY GRANDMA'S FREEZER BOX IN COMPARISON.

BUT LET'S GET BACK TO THE GIANT DIPROTODONS. THESE GUYS WERE TOUGH COOKIES. THEY'D BEEN AROUND IN AUSTRALIA FOR 1.5 MILLION YEARS. THEY MADE IT THROUGH AT LEAST TEN ICE AGES! AND THEY ALSO MADE IT THROUGH THE FIRST NUTSO PHASE OF THE LAST ICE AGE AROUND 70,000 YEARS AGO!

SO WHY WOULD THEY SUDDENLY CALL IT A DAY 50,000 YEARS AGO, HUH? EXACTLY WHEN YOUR CLIENTS MADE THEIR ENTRANCE?

PAH! THAT'S CIRCUMSTANTIAL AND JUST A COINCIDENCE...

NO, IT ISN'T, ADAMSKI.

IF THE DIPROTODONS WERE THE ONLY BIG GUYS TO DIE OUT THEN, SURE, YOU COULD SAY IT WAS A COINCIDENCE. BUT MORE THAN 90% OF AUSTRALIA'S MEGAFAUNA BOUGHT IT AT THE SAME TIME!

OK, SO IT'S CIRCUMSTANTIAL EVIDENCE... BUT GIVE ME A BREAK, DO YOU REALLY THINK IT'S A COINCIDENCE THAT YOUR CLIENTS ARRIVED IN AUSTRALIA AT THE EXACT SAME TIME THAT ALL THESE CRITTERS WERE DYING OF COLD?!

WELL, WHY NOT? COINCIDENCES DO HAPPEN. IS THAT ALL YOU GOT? YOU KNOW AS WELL AS I DO THAT THE COURT WILL FIND IN OUR FAVOR!

YOU'D HAVE THOUGHT, RIGHT? BUT I HAVE MORE EVIDENCE! WHEN CLIMATE CHANGE CAUSES MASS EXTINCTION, SEA CREATURES USUALLY TAKE JUST AS BIG A HIT AS THEIR COUSINS ON LAND, OK? WELL, THERE'S NO SIGN THAT ANY OCEAN ANIMALS AROUND AUSTRALIA VANISHED 50,000 YEARS AGO!

EVEN THOUGH THEIR LITTLE OCEAN-GOING HOBBY WAS HAVING A BOOM, YOUR CLIENTS WERE REALLY ONLY A THREAT ON LAND. THE STUFF THEY DID TOTALLY EXPLAINS WHY THIS WAVE OF EXTINCTION WIPED OUT AUSTRALIA'S LAND ANIMALS BUT DIDN'T TOUCH THE ONES IN THE OCEANS ALL AROUND!

IF I MAY, I'D LIKE TO ADD A THIRD ARGUMENT. THE ENSUING MILLENNIA SAW MASS EXTINCTIONS SIMILAR TO THIS ARCHETYPAL AUSTRALIAN DECIMATION WHENEVER HUMANS SETTLED ANOTHER PART OF THE OUTER WORLD.

YOU SEE, ADAMSKI, WHICHEVER WAY YOU LOOK AT IT, SAPIENS ARE GUILTY— IRREFUTABLY!

TAKE NEW ZEALAND FOR EXAMPLE. THE MEGAFAUNA THERE HAD WEATHERED THE SO-CALLED "CATASTROPHIC CLIMATE CHANGE" OF 50,000 YEARS AGO WITHOUT A SCRATCH, BUT THEN SUFFERED DEVASTATING BLOWS IMMEDIATELY AFTER THE FIRST HUMANS SET FOOT ON THE ISLANDS.

THE MĀORIS, NEW ZEALAND'S FIRST SAPIENS COLONIZERS, REACHED THE ISLANDS ABOUT 800 YEARS AGO. WITHIN A COUPLE OF CENTURIES, THE MAJORITY OF THE LOCAL MEGAFAUNA WAS EXTINCT.

...ALONG WITH 60% OF THE ISLANDS' BIRD SPECIES.

SOMETHING SIMILAR HAPPENED TO THE MAMMOTHS OF WRANGEL: AN ISLAND IN THE ARCTIC OCEAN, 125 MILES NORTH OF THE SIBERIAN COAST.

MAMMOTHS HAD FLOURISHED FOR MILLIONS OF YEARS OVER MOST OF THE NORTHERN HEMISPHERE BUT AS HOMO SAPIENS SPREAD ACROSS EURASIA AND THEN NORTH AMERICA, THE MAMMOTHS RETREATED.

BY ABOUT 10,000 YEARS AGO THERE WERE NO MAMMOTHS TO BE FOUND ANYWHERE, EXCEPT ON A FEW REMOTE ARCTIC ISLANDS, ESPECIALLY WRANGEL.

ARCTIC OCEAN

SIBERIA

WRANGEL ISLAND

WRANGEL'S MAMMOTHS DID WELL FOR A FEW THOUSAND YEARS MORE, THEN SUDDENLY DISAPPEARED ABOUT 4,000 YEARS AGO, JUST WHEN THE FIRST HUMANS ARRIVED.

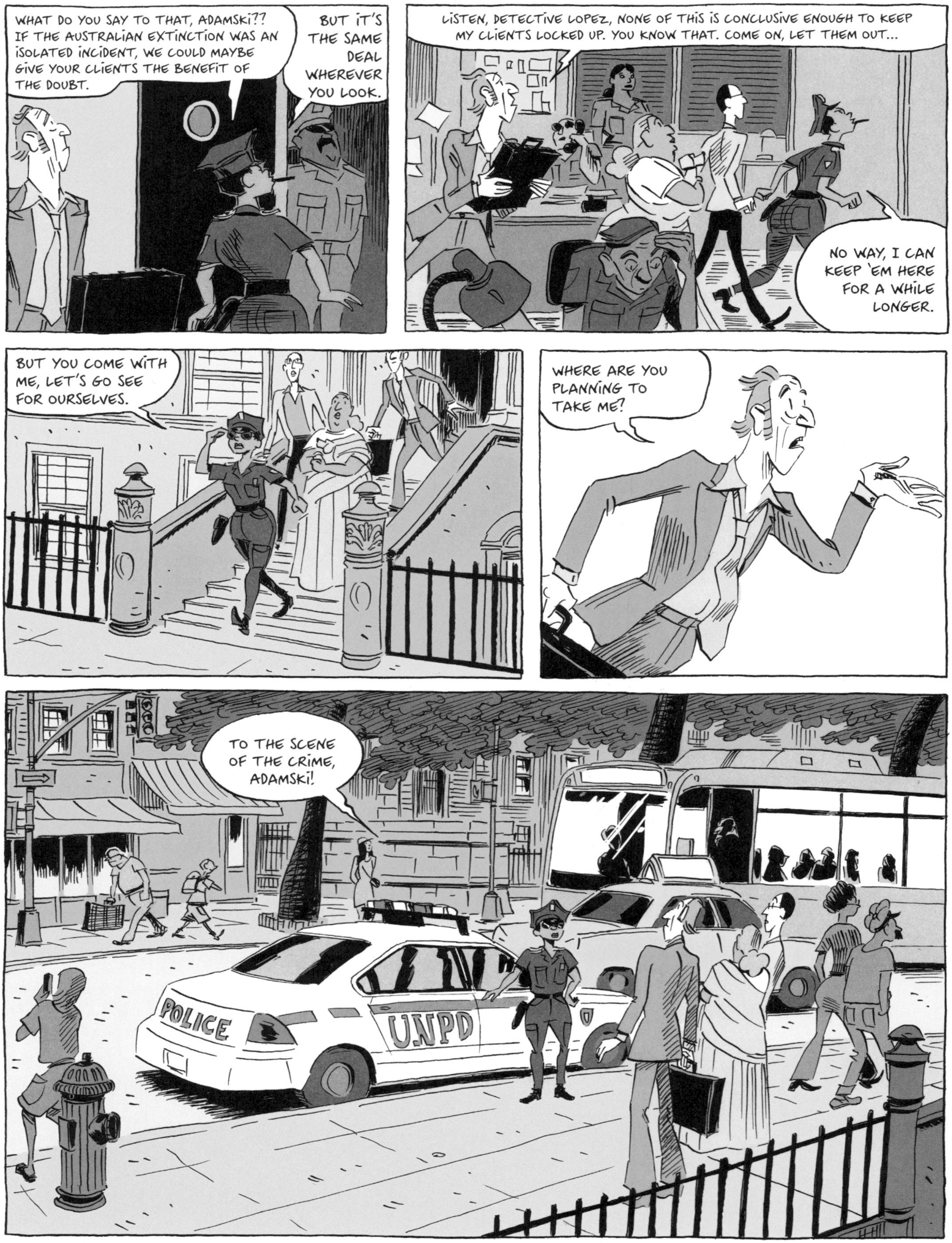
WHAT DO YOU SAY TO THAT, ADAMSKI?? IF THE AUSTRALIAN EXTINCTION WAS AN ISOLATED INCIDENT, WE COULD MAYBE GIVE YOUR CLIENTS THE BENEFIT OF THE DOUBT.
BUT IT'S THE SAME DEAL WHEREVER YOU LOOK.
LISTEN, DETECTIVE LOPEZ, NONE OF THIS IS CONCLUSIVE ENOUGH TO KEEP MY CLIENTS LOCKED UP. YOU KNOW THAT. COME ON, LET THEM OUT...
NO WAY, I CAN KEEP 'EM HERE FOR A WHILE LONGER.
BUT YOU COME WITH ME, LET'S GO SEE FOR OURSELVES.
WHERE ARE YOU PLANNING TO TAKE ME?
TO THE SCENE OF THE CRIME, ADAMSKI!
POLICE
UNPD

WHY THE HELL DO YOU SUSPECT MY CLIENTS OF DAMAGING AUSTRALIA? EVERYTHING LOOKS FINE TO ME. WOULD YOU LOOK AT THAT SCENERY!

OK, LOPEZ, LET'S ACCEPT THAT MY CLIENTS AND THEIR FRIENDS PASSED THROUGH HERE AT SOME POINT. BUT HOW CAN YOU BELIEVE THEY CAUSED AN ECOLOGICAL CATASTROPHE ALL ON THEIR OWN? THEY DIDN'T HAVE ATOM BOMBS, YOU KNOW! ALL THEY HAD WAS STONE-AGE TECHNOLOGY, FOR PETE'S SAKE!

WHERE'S YOUR IRREFUTABLE EVIDENCE, LOPEZ?

IT IS POSSIBLE—BUT THIS IS HYPOTHETICAL, OF COURSE, I DON'T ADMIT ANYTHING!—THAT THEY SOMETIMES HUNTED A DIPROTODON, MAYBE EVEN TWO. THEN WHAT? IT TAKES MORE THAN THAT TO CAUSE A MASS EXTINCTION!

I COULD ANSWER THAT, MR. ADAMSKI. THE LARGE ANIMALS THAT WERE THE PRIMARY VICTIMS OF THE AUSTRALIAN EXTINCTION BRED SLOWLY. THEY HAD LONG PREGNANCIES, FEW YOUNG PER PREGNANCY, AND LONG BREAKS IN BETWEEN.

SO, EVEN IF HUMANS KILLED ONLY ONE DIPROTODON EVERY TWO OR THREE MONTHS, THAT WOULD BE ENOUGH FOR DEATHS TO OUTNUMBER BIRTHS.

A FEW THOUSAND YEARS LATER THE LAST LONESOME DIPROTODON WOULD PASS AWAY, AND THE SPECIES WOULD BE GONE FOREVER.

WITH ALL DUE RESPECT, PROFESSOR, THAT'S NOT VERY CONVINCING. HUMANS FIRST EVOLVED IN AFRICA AND THEY STARTED HUNTING LARGE ANIMALS THERE BEFORE HEADING OUT TO SETTLE IN OTHER PLACES.

BUT THERE ARE STILL ELEPHANTS, RHINOS AND HIPPOS IN AFRICA! IN FACT, AFRICA'S THE PLACE WITH THE BIGGEST ANIMALS! HOW DO YOU EXPLAIN THAT, LOPEZ?!

NOT SO FAST, ADAMSKI. YOU'RE FORGETTING EVOLUTION.

THAT'S RIGHT! WHEN HUMANS STARTED HUNTING IN AFRICA, THEY WERE STILL TOTAL ROOKIES. WHILE THEY SLOWLY GOT THEIR ACT TOGETHER, THE BIG ANIMALS HAD TIME TO LEARN TO BE SCARED OF 'EM. BY THE TIME SAPIENS WAS THE DEADLIEST SPECIES ON EARTH, AFRICA'S BIG ANIMALS ALREADY KNEW THEY DID NOT WANT TO HANG OUT WITH THESE TWO-LEGGED APES.

SO, BASICALLY YOU'RE SAYING AFRICAN ANIMALS WERE SMART AND RAN AWAY BUT AUSTRALIAN ANIMALS WERE SO DUMB THEY JUST LET THEMSELVES BE SLAUGHTERED?

IT'S NOT ABOUT INTELLIGENCE, IT'S ABOUT EXPERIENCE! IT WAS WAY EASIER TO HUNT THE BIG GUYS IN AUSTRALIA THAN IN AFRICA, BECAUSE THE AUSTRALIAN ANIMALS HADN'T LEARNED TO BE SCARED OF HUMANS!

YOU WANNA KNOW SOMETHING, ADAMSKI? YOUR CLIENTS DON'T EXACTLY LOOK SUPER DANGEROUS. THEY DON'T HAVE LONG SHARP TEETH, THEY'RE NOT AGILE AND RIPPED...

DIPROTODONS WERE THE LARGEST MARSUPIALS EVER TO WALK THE EARTH, AND WHEN THEY FIRST SAW THESE FRAIL-LOOKING APES, THIS IS WHAT PROBABLY HAPPENED...

LEARNING TO BE AFRAID OF NEW DANGERS TAKES TIME.

THE CLASSIC EXAMPLE IS THE POOR DODO...

DODO? I DIDN'T SEE THAT NAME IN THE FILE...

I MEAN THE EXTINCT BIRD, MR. ADAMSKI. DODOS LIVED ON MAURITIUS FOR MILLIONS OF YEARS.

THERE WERE NO PREDATORS IN THIS ISOLATED PLACE, SO THESE BIRDS HAD NO FEAR AS WELL AS NO WINGS. WHEN DUTCH SAILORS DISCOVERED MAURITIUS, THE DODOS DIDN'T TRY TO RUN AWAY. THE SAILORS COULD KILL AS MANY AS THEY LIKED. IT WASN'T LONG BEFORE DODOS WERE EXTINCT.

THINK ABOUT THE PHOBIAS THAT WE HUMANS HAVE. LOTS OF PEOPLE ARE FRIGHTENED OF SNAKES OR SPIDERS, BUT NO ONE'S SCARED OF CARS EVEN THOUGH THEY KILL MANY MORE PEOPLE AROUND THE WORLD THAN SNAKES OR SPIDERS DO!

WHY IS THAT? BECAUSE THE ONE CENTURY THAT'S ELAPSED SINCE ARMAND PEUGEOT OPENED HIS CAR FACTORY ISN'T ENOUGH FOR US TO HAVE DEVELOPED A HEALTHY FEAR OF CARS. CATS ARE A GOOD EXAMPLE. WHEN A CAR COMES STRAIGHT AT A CAT, THE ANIMAL OFTEN STANDS THERE, PARALYZED. IS THE CAT STUPID? NO. BUT ITS INSTINCTS HAVEN'T HAD TIME TO ADAPT TO DEALING WITH CARS.

IN EXACTLY THE SAME WAY, DIPROTODONS DIDN'T HAVE A CHANCE TO DEVELOP A HEALTHY FEAR OF HUMANS. LIKE THOSE POOR DODOS, THEY PROBABLY ALL DIED BEFORE THEY EVEN UNDERSTOOD WHAT WAS ATTACKING THEM.

AND BY THE TIME YOUR CLIENTS REACHED AUSTRALIA, THEY'D NOT ONLY MASTERED HUNTING BUT ALSO LEARNED TO USE FIRE. FACED WITH UNFAMILIAR AND THREATENING FORESTS THEY COULD DELIBERATELY BURN THEM DOWN. THIS CREATED OPEN GRASSLAND, WHICH ATTRACTED GAME THAT WAS EASIER TO HUNT.

WITH THESE TACTICS THEY COMPLETELY CHANGED THE ECOLOGY OF LARGE PARTS OF AUSTRALIA IN JUST A FEW SHORT MILLENNIA.

I'VE READ THE EVIDENCE CAREFULLY, PROFESSOR! GRANTED, A LOT OF TREES AND SCRUB DISAPPEARED AS A RESULT OF MY CLIENTS' FIRES...

BUT WHAT ABOUT EUCALYPTUS TREES? THEY WERE RARE IN AUSTRALIA 50,000 YEARS AGO AND THEY STARTED SPREADING VERY QUICKLY! THEY RECOVERED BETTER AFTER FIRES. SO MY CLIENTS' ARRIVAL USHERED IN A GOLDEN AGE FOR THE EUCALYPTUS! AND THERE'S SUPPORTING EVIDENCE! LOOK AT THESE PHOTOS OF FOSSILIZED PLANTS! EVERYTHING CORROBORATES IT!

AND THAT'S NOT ALL! THESE CHANGES IN VEGETATION INFLUENCED THE ANIMALS THAT ATE THE PLANTS AND THE CARNIVORES THAT ATE THOSE ANIMALS. TAKE KOALAS, FOR EXAMPLE. THEY EAT NOTHING BUT EUCALYPTUS LEAVES SO THEY HAPPILY MUNCHED THEIR WAY INTO NEW TERRITORIES!! AND ALL THANKS TO MY CLIENTS!

AND WHAT COULD BE CUTER THAN A PRETTY LITTLE KOALA, TELL ME THAT, HMM?

LOPEZ, YOU NEED TO THINK HOW THE JURY WILL REACT WHEN I CALL CUTE KOALAS AS WITNESSES.

FREE BILL & CINDY

YES, BUT MOST OTHER ANIMALS SUFFERED TERRIBLY, MR. ADAMSKI. MANY AUSTRALIAN FOOD CHAINS COLLAPSED, DRIVING THE WEAKEST LINKS INTO EXTINCTION.

I'M NOT CONVINCED. I STILL MAINTAIN THAT IT WAS ALL BECAUSE OF CLIMATE CHANGE.

HIS ALLEGATION ISN'T ENTIRELY FALSE, YOU KNOW. THE CLIMATE CHANGES THAT HIT AUSTRALIA ABOUT 50,000 YEARS AGO REALLY DID DESTABILIZE THE ECOSYSTEM AND MADE IT PARTICULARLY VULNERABLE!

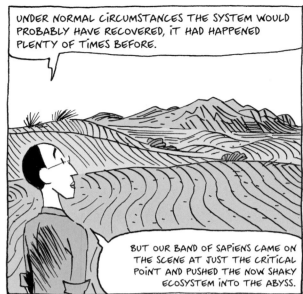

UNDER NORMAL CIRCUMSTANCES THE SYSTEM WOULD PROBABLY HAVE RECOVERED, IT HAD HAPPENED PLENTY OF TIMES BEFORE.

BUT OUR BAND OF SAPIENS CAME ON THE SCENE AT JUST THE CRITICAL POINT AND PUSHED THE NOW SHAKY ECOSYSTEM INTO THE ABYSS.

THIS COMBINATION OF CLIMATE CHANGE AND HUMAN HUNTING HIT LARGE ANIMALS ESPECIALLY HARD BECAUSE IT ATTACKED THEM FROM DIFFERENT ANGLES.

YES, IT'S HARD TO FIND A GOOD SURVIVAL STRATEGY THAT WORKS AGAINST SEVERAL THREATS AT THE SAME TIME...

WITHOUT MORE EVIDENCE, IT'S HARD TO SAY WHAT PROPORTION OF BLAME LIES WITH CLIMATE CHANGE ON THE ONE HAND AND YOUR CLIENTS ON THE OTHER.

BUT WE STILL HAVE A BUNCH OF REASONS TO SUGGEST THAT IF YOUR CLIENTS NEVER SET FOOT IN AUSTRALIA, THE PLACE WOULD STILL HAVE MARSUPIAL LIONS, DIPROTODONS AND GIANT KANGAROOS!

C'MON, LET'S GET BACK TO THE PLANE. WE'VE GOT SOME MORE CRIME SCENES TO VISIT.

SIBERIA

AUSTRALIA

WHERE ARE YOU TAKING US, LOPEZ?

FOLLOWING IN YOUR CLIENTS' FOOTSTEPS!

THE EXTINCTION OF THE AUSTRALIAN MEGAFAUNA WAS JUST THE BEGINNING, ADAMSKI! JUST A REHEARSAL!

YOUR CLIENTS PERPETRATED AN EVEN BIGGER ECOLOGICAL CATASTROPHE IN AMERICA!

EXCUSE ME! ARE YOU INSINUATING THAT MY CLIENTS WERE ALSO ACTIVE IN THE AMERICAS???

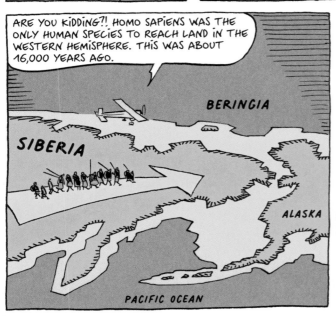

ARE YOU KIDDING?! HOMO SAPIENS WAS THE ONLY HUMAN SPECIES TO REACH LAND IN THE WESTERN HEMISPHERE. THIS WAS ABOUT 16,000 YEARS AGO.

BERINGIA

SIBERIA

ALASKA

PACIFIC OCEAN

WOULD YOU JUST LOOK AT THE PLACE BACK THEN! SEA LEVELS WERE SO LOW THAT THERE WAS A LAND BRIDGE FROM NORTHEASTERN SIBERIA TO NORTHWESTERN ALASKA! SO THE FIRST HUMANS TO REACH AMERICA FROM SIBERIA EITHER STROLLED RIGHT IN OR CRUISED ALONG THE COAST IN LITTLE BOATS.

NOT THAT IT WAS EASY! THE JOURNEY WAS TOUGH, PERHAPS HARDER THAN THE SEA CROSSING TO AUSTRALIA. AND FIRST, SAPIENS HAD TO WITHSTAND NORTHERN SIBERIA'S EXTREME ARCTIC CONDITIONS. THE SUN DOESN'T COME UP AT ALL IN WINTER AND TEMPERATURES CAN DROP TO −60°F.

COULD WE CLOSE THE WINDOW?

OH, COME ON! ARE YOU SURE IT WAS MY CLIENTS? DID YOU TAKE A GOOD LOOK AT THEM? SAPIENS WERE CUT OUT FOR LIFE ON THE HOT AFRICAN SAVANNA, NOT SIBERIA! EVEN THE NEANDERTHALS, WHO WERE ADAPTED TO COLD CLIMATES, NEVER SUCCEEDED IN COLONIZING NORTHERN SIBERIA.

YOU'RE RIGHT, YOUR CLIENTS' BODIES WERE ADAPTED TO LIFE ON THE AFRICAN SAVANNA, BUT THEIR MINDS FOUND INGENIOUS SOLUTIONS TO SURVIVE IN LANDS OF SNOW AND ICE!

WHEN BANDS OF FORAGERS MIGRATED INTO COLDER CLIMATES, THEY LEARNED TO MAKE SNOWSHOES AND THERMAL CLOTHING USING LAYERS OF FURS AND SKINS TIGHTLY SEWN TOGETHER WITH A NEEDLE.

218

THEY DEVELOPED NEW WEAPONS AND COOPERATIVE HUNTING TECHNIQUES SO THEY COULD TRACK AND KILL MAMMOTHS AND OTHER BIG GAME IN THE FAR NORTH.

AS THEIR THERMAL CLOTHING AND HUNTING TECHNIQUES IMPROVED, SAPIENS DARED TO VENTURE DEEPER INTO THESE FROZEN REGIONS. AND AS THEY MOVED NORTH, THEIR CLOTHES, HUNTING STRATEGIES AND OTHER SURVIVAL SKILLS KEPT ON GETTING BETTER.

BUT JUST LOOK AT THIS HOSTILE PLACE! IT DOESN'T MAKE ANY SENSE! WHY WOULD ANYONE CHOOSE TO BANISH THEMSELVES TO SIBERIA?

MAYBE THE SUCKERS WERE DRIVEN NORTH BY WAR, ADAMSKI!!!

OR MAYBE THERE WERE DEMOGRAPHIC PRESSURES OR NATURAL CATASTROPHES....

BUT SOME OF THEM MAY HAVE BEEN LURED BY MORE POSITIVE REASONS... I MEAN THESE FREEZING EXPANSES WERE FULL OF BIG, JUICY ANIMALS LIKE REINDEER AND MAMMOTHS.

EVERY MAMMOTH WAS A HUGE SOURCE OF MEAT—AND IN THOSE TEMPERATURES, THEY COULD EVEN FREEZE IT FOR LATER! BUT IT WAS ALSO A SOURCE OF TASTY FAT, WARM FUR AND PRECIOUS IVORY.

WOOF! WOOF! WOOF!

REMEMBER SUNGIR? THE ARCHAEOLOGICAL EVIDENCE FROM THAT SITE PROVES THAT MAMMOTH-HUNTERS DIDN'T JUST SURVIVE IN THE FROZEN NORTH—THEY THRIVED. THEY GRADUALLY SPREAD FAR AND WIDE, FOLLOWING THE MAMMOTHS, MASTODONS, RHINOCEROSES AND REINDEER.

WOOF! WOOF! WOOF!

YEAH, AND PLUS DRIVING THESE POOR CRITTERS TO EXTINCTION...

WOOF!

ENOUGH, KIKI! SORRY ABOUT HER! WE SHOULD GET BACK TO THE PLANE.

WOOF! WOOF!

AROUND 16,000 YEARS AGO, THEIR HUNTING TOOK SOME OF THEM FROM NORTHEASTERN SIBERIA INTO ALASKA.

SIBERIA

ALASKA

OF COURSE, THEY DIDN'T REALIZE THEY WERE DISCOVERING A NEW WORLD! TO THEM—AND TO THE MAMMOTHS— ALASKA WAS JUST AN EXTENSION OF SIBERIA.

AT FIRST, THERE WERE GLACIERS BLOCKING OFF ALASKA FROM THE REST OF AMERICA SO ONLY A HANDFUL OF PIONEERS MADE IT THROUGH TO EXPLORE FURTHER SOUTH.

THEN ABOUT 14,000 YEARS AGO, GLOBAL WARMING MELTED THE ICE, MAKING THE JOURNEY MUCH EASIER.

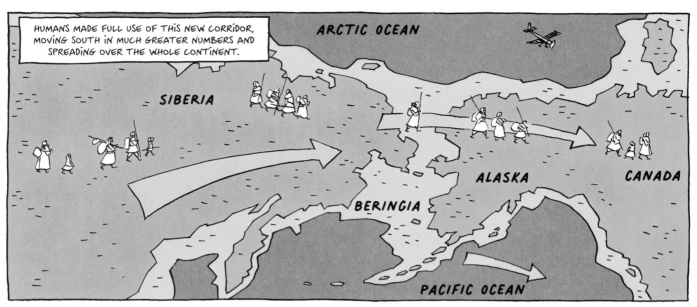

HUMANS MADE FULL USE OF THIS NEW CORRIDOR, MOVING SOUTH IN MUCH GREATER NUMBERS AND SPREADING OVER THE WHOLE CONTINENT.

ARCTIC OCEAN

SIBERIA

BERINGIA

ALASKA

CANADA

PACIFIC OCEAN

OF COURSE, THEY WERE ORIGINALLY ADAPTED TO HUNTING LARGE GAME IN THE ARCTIC, BUT THEY SOON ADJUSTED TO AN AMAZING VARIETY OF CLIMATES AND ECOSYSTEMS.

DESCENDANTS OF THE SIBERIANS COLONIZED THE THICK FORESTS OF THE EASTERN UNITED STATES...

THE SWAMPS OF THE MISSISSIPPI DELTA...

THE DESERTS OF MEXICO...

AND THE STEAMING JUNGLES OF CENTRAL AMERICA...

SOME SETTLED IN THE AMAZON BASIN...

SOME IN THE OPEN PAMPAS OF ARGENTINA...

OTHERS ALL THE WAY DOWN IN TIERRA DEL FUEGO...

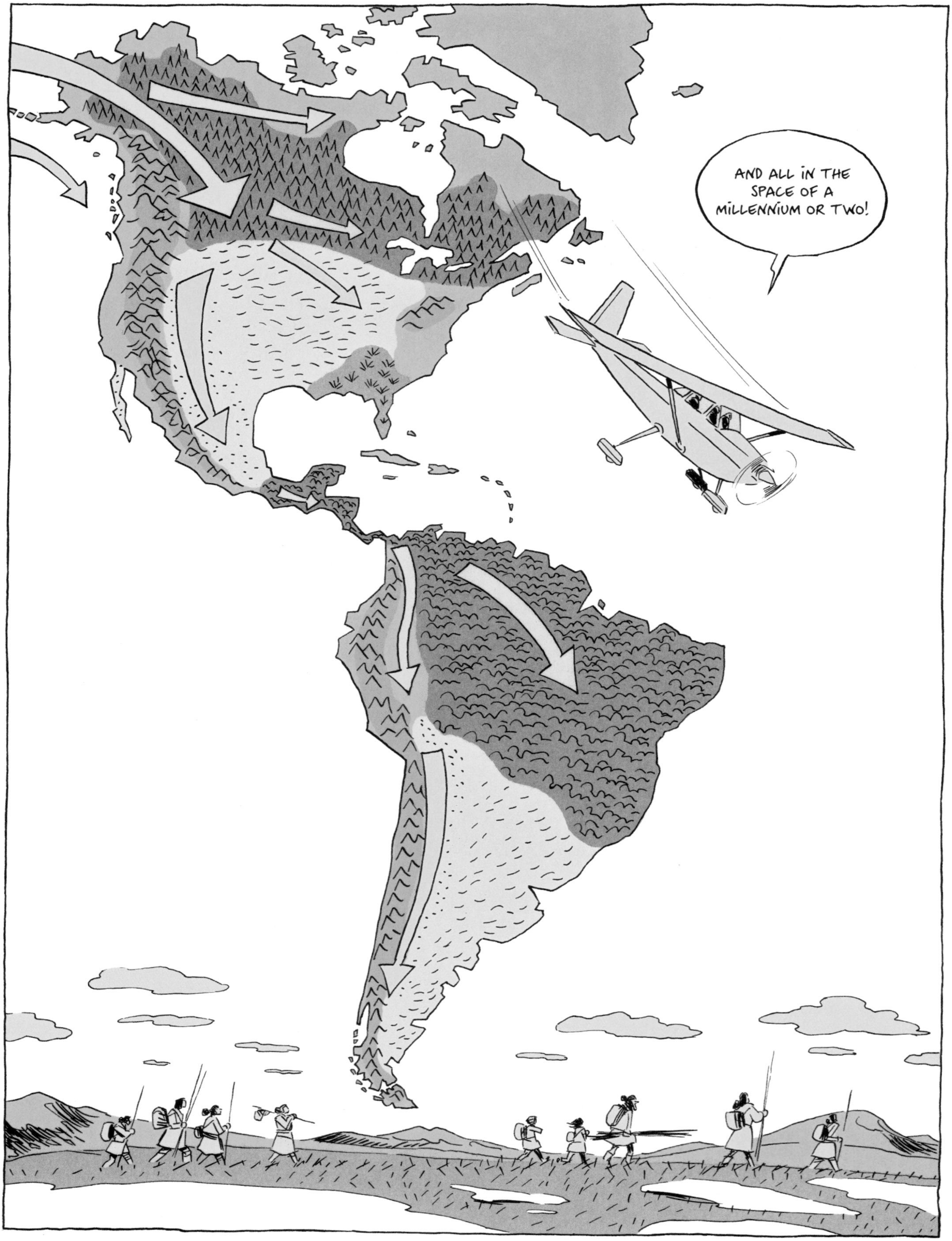
AND ALL IN THE SPACE OF A MILLENNIUM OR TWO!

BY ABOUT 12,000 YEARS AGO, HUMANS ALREADY INHABITED THE SOUTHERNMOST TIP OF SOUTH AMERICA, THE ISLAND OF TIERRA DEL FUEGO.

YEAH, AND YOU KNOW WHAT THIS STORMING OF THE AMERICAN CONTINENT SHOWS? IT SHOWS JUST HOW INGENIOUS AND ADAPTABLE YOUR CLIENTS ARE!

WE'RE ABOUT TO LAND, GUYS, BUCKLE UP!

NO OTHER ANIMAL HAD EVER SPREAD SO QUICKLY INTO SUCH A VARIETY OF RADICALLY DIFFERENT HABITATS—AND THEY MANAGED IT WITH VIRTUALLY THE SAME GENES THROUGHOUT.

HMM, AND LEAVING A LONG TRAIL OF VICTIMS BEHIND 'EM...

AMERICAN FAUNA WAS FAR RICHER 14,000 YEARS AGO THAN IT IS TODAY. WHEN THE FIRST AMERICANS HEADED SOUTH FROM ALASKA INTO THE PLAINS OF CANADA AND THE WESTERN UNITED STATES, THEY FOUND MAMMOTHS AND MASTODONS...

RODENTS THE SIZE OF BEARS...

HERDS OF HORSES AND CAMELS...

GIANT LIONS...

AND DOZENS OF SPECIES THAT DON'T EVEN EXIST TODAY, INCLUDING TERRIFYING SABER-TOOTH CATS...

AND GIANT GROUND SLOTHS THAT WEIGHED UP TO EIGHT TONS AND COULD BE 20 FEET IN HEIGHT!

THERE WAS AN EVEN MORE EXOTIC MENAGERIE OF LARGE MAMMALS, REPTILES AND BIRDS IN SOUTH AMERICA. YOU COULD SAY THE AMERICAS WERE A HUGE LABORATORY OF EVOLUTIONARY EXPERIMENTATION, WHERE MANY ANIMALS AND PLANTS THAT DIDN'T EXIST IN AFRICA OR ASIA HAD EVOLVED AND THRIVED.

BUT NOT ANYMORE. WITHIN 2,000 YEARS OF SAPIENS ARRIVING, MOST OF THESE UNIQUE SPECIES HAD BEEN WIPED OUT. SCIENTISTS ESTIMATE THAT, IN THAT SHORT TIME, NORTH AMERICA LOST 34 OF ITS 47 GENERA OF LARGE MAMMALS, AND SOUTH AMERICA LOST 50 OUT OF 60.

THE SABER-TOOTH CATS, WHICH HAD FLOURISHED FOR 30 MILLION YEARS, DISAPPEARED.

SO DID THE GIANT GROUND SLOTHS, THE HUGE LIONS, THE NATIVE AMERICAN HORSES AND CAMELS, THE GIANT RODENTS AND THE MAMMOTHS.

THOUSANDS OF SPECIES OF SMALLER MAMMALS, REPTILES, BIRDS AND EVEN INSECTS AND PARASITES BECAME EXTINCT TOO. WHEN ALL THE MAMMOTHS DIED—ALL THE SPECIES OF MAMMOTH TICKS FOLLOWED THEIR GIANT HOSTS INTO OBLIVION!

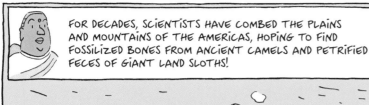

FOR DECADES, SCIENTISTS HAVE COMBED THE PLAINS AND MOUNTAINS OF THE AMERICAS, HOPING TO FIND FOSSILIZED BONES FROM ANCIENT CAMELS AND PETRIFIED FECES OF GIANT LAND SLOTHS!

WHEN THEY FIND WHAT THEY'RE LOOKING FOR, THEY CAREFULLY WRAP THEIR TREASURES AND SEND THEM OFF TO LABS WHERE EVERY BONE AND EVERY FOSSILIZED POOP IS METICULOUSLY STUDIED AND DATED!

AND ALL THE ANALYSIS JUST KEEPS PRODUCING THE SAME RESULTS: THE MOST RECENT CAMEL BONES AND DUNG BALLS DATE TO THE TIME WHEN HUMANS FLOODED ACROSS AMERICA, BETWEEN ABOUT 14,000 AND 11,000 YEARS AGO.

THERE'S ONLY ONE AREA WHERE RESEARCHERS HAVE FOUND MORE RECENT GROUND SLOTH DUNG BALLS—THE CARIBBEAN. ON SOME ISLANDS, PARTICULARLY CUBA AND HISPANIOLA, THEY FOUND PETRIFIED SLOTH DROPPINGS THAT DATED TO ABOUT 7,000 YEARS AGO.

AND THAT'S EXACTLY WHEN THE FIRST HUMANS MANAGED TO CROSS THE CARIBBEAN SEA AND SETTLED IN THESE TWO ISLANDS.

WITH ALL DUE RESPECT, PROFESSOR, I DON'T THINK YOU'RE GOING TO TEAR APART MY DEFENSE WITH ALL THAT SH...

YOU KNOW WHAT, ADAMSKI, IN AMERICA NOBODY CAN DODGE THE—AH—DUNG BALL!

THERE'S NO WAY AROUND THE FACTS. CLIMATE CHANGE MAY HAVE GIVEN YOUR CLIENTS A HELPING HAND, BUT THEIR CONTRIBUTION TO THE ECOLOGICAL DEVASTATION OF AMERICA WAS DECISIVE.

WE'VE GOT ALL THE EVIDENCE WE NEED. LET'S DO THIS!

NEW YORK, THREE MONTHS LATER.

IT'S GREAT YOU'RE HERE, ZOE, WE'RE GOING TO WITNESS A HISTORIC EVENT.

TODAY'S THE MOST IMPORTANT DAY IN THE SAPIENS' TRIAL. WE'RE FINALLY GOING TO HEAR THE COURT'S VERDICT.

THAT'S SO SAD! THEY'RE IN SUCH BIG TROUBLE! THEY REALLY DON'T LOOK SO HORRIBLE.

DON'T YOU WORRY, THEY HAVE A LAWYER.
BUT THE EVIDENCE FEELS SO—WHAT'S THE WORD?— OVERWHELMING...

AND THEIR ADAMSKI DOESN'T LOOK ALL THAT SMART...

DON'T BE FOOLED BY APPEARANCES... COME ON, LET'S GO IN, IT'S ABOUT TO START.

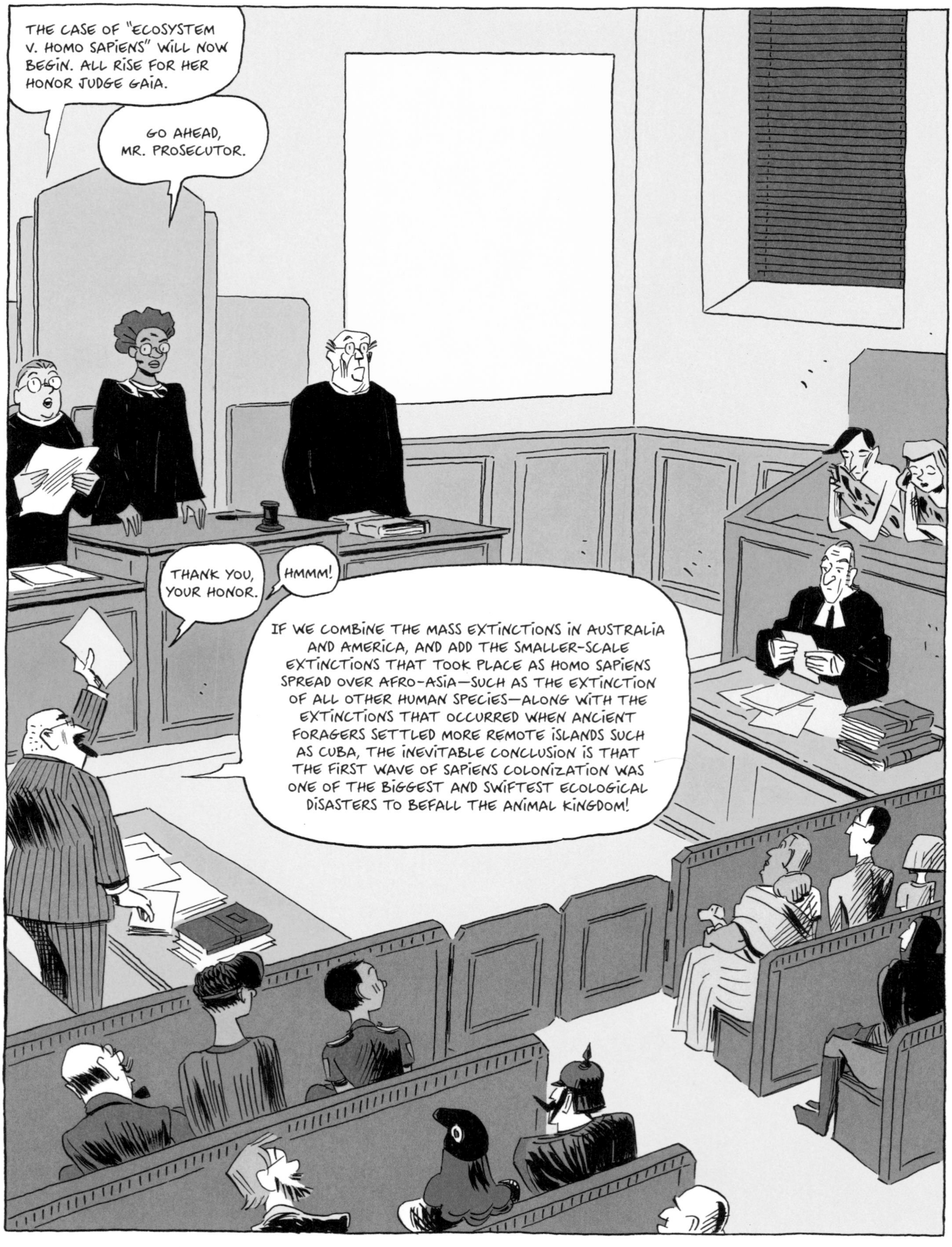
THE CASE OF "ECOSYSTEM V. HOMO SAPIENS" WILL NOW BEGIN. ALL RISE FOR HER HONOR JUDGE GAIA.
GO AHEAD, MR. PROSECUTOR.
THANK YOU, YOUR HONOR.
HMMM!
IF WE COMBINE THE MASS EXTINCTIONS IN AUSTRALIA AND AMERICA, AND ADD THE SMALLER-SCALE EXTINCTIONS THAT TOOK PLACE AS HOMO SAPIENS SPREAD OVER AFRO-ASIA—SUCH AS THE EXTINCTION OF ALL OTHER HUMAN SPECIES—ALONG WITH THE EXTINCTIONS THAT OCCURRED WHEN ANCIENT FORAGERS SETTLED MORE REMOTE ISLANDS SUCH AS CUBA, THE INEVITABLE CONCLUSION IS THAT THE FIRST WAVE OF SAPIENS COLONIZATION WAS ONE OF THE BIGGEST AND SWIFTEST ECOLOGICAL DISASTERS TO BEFALL THE ANIMAL KINGDOM!

HARDEST HIT WERE LARGE FUR-COVERED CREATURES. AT THE TIME OF THE COGNITIVE REVOLUTION, THE PLANET WAS HOME TO SOME 200 GENERA OF TERRESTRIAL MAMMALS THAT WEIGHED MORE THAN 100 POUNDS.

BY THE TIME OF THE AGRICULTURAL REVOLUTION, ONLY ABOUT 100 REMAINED. THE ACCUSED BROUGHT ABOUT THE EXTINCTION OF NEARLY HALF THE PLANET'S LARGE ANIMALS LONG BEFORE HUMANS INVENTED THE WHEEL, WRITING OR IRON TOOLS!

BUT, LADIES AND GENTLEMEN OF THE JURY, THERE'S WORSE! BECAUSE THE ACCUSED CONTINUED WITH THEIR KILLING SPREE LONG AFTER THAT!!!

AFTER THE AGRICULTURAL REVOLUTION, THE ACCUSED DIVERSIFIED INTO FARMING. BUT EVEN AS FARMERS, THEY CONTINUED TO COLONIZE NEW TERRITORIES AND TO DRIVE THE INDIGENOUS FAUNA TO EXTINCTION.

AND THE ACCUSED USED THE SAME MODUS OPERANDI EVERY TIME! ARCHAEOLOGICAL FINDS ON ISLAND AFTER ISLAND TELL THE SAME SAD STORY.

THE DEEPEST ARCHAEOLOGICAL STRATUM SHOWS TRACES OF A RICH AND VARIED POPULATION OF LARGE ANIMALS BUT NO TRACE OF HUMAN PRESENCE.

AND THEN COME THE FIRST SIGNS OF THE SAPIENS GANG AT THE CRIME SCENE: A HUMAN BONE, A SPEARHEAD, A FRAGMENT OF POTTERY OR PERHAPS A HUMAN TOOTH.

AFTER THAT IT'S ALWAYS THE SAME SCENARIO: THE ISLAND FILLS UP WITH HUMAN HOUSES AND FARMS WHILE THE LARGE ANIMALS—AND A GOOD MANY OF THE SMALLER ONES—ARE WIPED OUT!

WE HAVE A FAMOUS EXAMPLE IN THE LARGE ISLAND OF MADAGASCAR, ABOUT 250 MILES EAST OF THE AFRICAN MAINLAND. THANKS TO MILLIONS OF YEARS OF ISOLATION, A UNIQUE COLLECTION OF ANIMALS EVOLVED THERE.

MADAGASCAR

LADIES AND GENTLEMEN OF THE JURY, I GIVE YOU THE ELEPHANT BIRD, A FLIGHTLESS CREATURE 10 FEET TALL AND WEIGHING ALMOST HALF A TON—WE ARE TALKING ABOUT THE BIGGEST BIRD IN THE WORLD!

AND GIANT LEMURS... YOU GUESSED IT: THE LARGEST PRIMATES ON THE PLANET!

ALONG WITH MOST OTHER LARGE ANIMALS IN MADAGASCAR, ELEPHANT BIRDS AND GIANT LEMURS SUDDENLY VANISHED 1,500 YEARS AGO—PRECISELY WHEN THE ACCUSED FIRST SET FOOT ON THE ISLAND.

LET'S MOVE TO THE PACIFIC, WHERE THE MAIN WAVE OF EXTINCTIONS BEGAN ABOUT 3,500 YEARS AGO, WHEN THE SAPIENS GANG SETTLED FIJI AND NEW CALEDONIA.

DIRECTLY OR INDIRECTLY, THEY KILLED OFF HUNDREDS OF SPECIES OF BIRDS, INSECTS, SNAILS AND OTHER LOCAL INHABITANTS.

FROM THERE, THE WAVE OF EXTINCTION MOVES GRADUALLY TO THE EAST, THE SOUTH AND THE NORTH, INTO THE HEART OF THE PACIFIC OCEAN, OBLITERATING THE UNIQUE FAUNA OF SAMOA AND TONGA 3,200 YEARS AGO...

...OF THE MARQUESAS ISLANDS 2,000 YEARS AGO...

...OF EASTER ISLAND...

...THE COOK ISLANDS...

...AND HAWAii, ALL AROUND 1,500 YEARS AGO...

ENDING UP IN NEW ZEALAND 800 YEARS AGO!

SIMILAR ECOLOGICAL DISASTERS OCCURRED ON ALMOST EVERY ONE OF THE THOUSANDS OF ISLANDS PEPPERED ACROSS THE ATLANTIC OCEAN, THE INDIAN OCEAN, THE ARCTIC OCEAN AND THE MEDITERRANEAN SEA.

EVEN ON THE TINIEST ISLANDS THERE'S ARCHAEOLOGICAL EVIDENCE OF BIRDS, INSECTS AND SNAILS THAT LIVED THERE FOR COUNTLESS GENERATIONS ONLY TO VANISH WHEN THE SAPIENS GANG ARRIVED.

THE ACCUSED SPARED ONLY A FEW EXTREMELY REMOTE ISLANDS UNTIL THE MODERN AGE, AND—THERE'S A PATTERN HERE, I'M SURE YOU'LL AGREE—THESE ISLANDS KEPT THEIR FAUNA INTACT. LET'S TAKE A FAMOUS EXAMPLE: THE GALÁPAGOS ISLANDS WERE UNINHABITED BY HUMANS UNTIL THE 19TH CENTURY, SO THEY KEPT THEIR UNIQUE MENAGERIE, INCLUDING GIANT TORTOISES, WHICH, LIKE ANCIENT DIPROTODONS, SHOW NO FEAR OF HUMANS!

I KNOW MR. ADAMSKI HERE WILL TRY TO CONVINCE YOU THAT THE ACCUSED WERE HARMLESS HUNTER-GATHERERS, WHO LIVED IN HARMONY WITH NATURE.

DON'T FALL INTO THAT TRAP! THESE SAPIENS WERE THE MOST MURDEROUS CREATURES IN THE HISTORY OF PLANET EARTH!

GIVEN THE ACCUMULATION OF DAMNING EVIDENCE AND THE DEVASTATING CATALOGUE OF CRIMES FOR WHICH WE HOLD THE ACCUSED HERE PRESENT RESPONSIBLE, I SUGGEST, LADIES AND GENTLEMEN OF THE JURY, THAT THEY DESERVE THE MAXIMUM SENTENCE!

OH NO!!

SILENCE! SILENCE IN COURT! THE DEFENSE MAY NOW SPEAK!

PLEASE, MR. ADAMSKI, GO AHEAD.

AHEM... THANK YOU, YOUR HONOR...

BUT BEFORE I START, MAY I ASK FOR THE COURT TO GRANT MY CLIENTS PERMISSION TO WITHDRAW FROM THE COURTROOM FOR FIVE MINUTES? I WILL CALL THEM BACK IN LATER.

REQUEST GRANTED, MR. ADAMSKI.

THANK YOU, YOUR HONOR.

I WON'T EVEN ATTEMPT TO CONTRADICT THE PROSECUTOR'S DAZZLING OPENING STATEMENT. THE EVIDENCE IS TOO OVERWHELMING...

THIS MAY SEEM A SURPRISING TACTIC, BUT I SHALL START MY DEFENSE BY AGREEING WHOLEHEARTEDLY WITH THE PROSECUTION'S CASE...

BUT I MUST ASK YOU TO TAKE INTO CONSIDERATION THE FACT THAT MY CLIENTS WERE UNAWARE OF THE CONSEQUENCES OF THEIR ACTIONS. THEY DID NOT INTENTIONALLY ERADICATE MAMMOTHS AND DIPROTODONS.

TRUE, THE EXTINCTION OF THESE LARGE ANIMALS WAS FAST ON AN EVOLUTIONARY SCALE, BUT IT WAS SLOW AND GRADUAL FROM THE HUMANS' POINT OF VIEW.

PEOPLE LIVED NO MORE THAN 70 OR 80 YEARS, BUT THIS PROCESS OF EXTINCTION TOOK CENTURIES.

MY CLIENTS DIDN'T MAKE THE CONNECTION BETWEEN THEIR ANNUAL MAMMOTH HUNT—WHEN ONLY TWO OR THREE MAMMOTHS WERE KILLED—AND THE WHOLESALE DISAPPEARANCE OF THESE PRECIOUS GIANTS.

AT THE VERY MOST, SOME ELDER MIGHT HAVE TOLD A GROUP OF SKEPTICAL YOUNGSTERS THAT "THERE WERE MANY MORE MAMMOTHS AROUND WHEN I WAS A BOY."

THE SAME GOES FOR THE MASTODONS AND GIANT DEER. AND, OF COURSE, TRIBAL LEADERS ALWAYS TOLD THE TRUTH, AND CHILDREN RESPECTED THEIR ELDERS BACK THEN!

NEVERTHELESS, IGNORANCE DOES NOT ENTIRELY EXONERATE MY CLIENTS.

I HAVE TO CONCEDE THAT THEY WERE PARTLY RESPONSIBLE FOR THE TWO WAVES OF EXTINCTIONS DESCRIBED BY THE PROSECUTOR.

BUT MY CLIENTS DID NOT ACT ALONE, YOU KNOW.

DETECTIVE LOPEZ TOLD ME A LONG TIME AGO THAT SHE SUSPECTED MY CLIENTS WERE PART OF A MUCH LARGER GANG THAT SHE HOPED TO UNCOVER.

WELL, SHE WAS ABSOLUTELY RIGHT.

IN FACT, THERE WERE OTHER INDIVIDUALS WITHIN THE SAPIENS GANG WHO ARE FAR MORE GUILTY THAN MY CLIENTS.

YOU SEE, AFTER THE HUNTER-GATHERERS TRIGGERED THE FIRST WAVE OF EXTINCTIONS AND THE FARMERS CAUSED THE SECOND, THERE WAS A THIRD WAVE!

OH YES, NO ONE'S SAID THIS YET, BUT THE KILLINGS CONTINUED JUST AS INTENSIVELY DURING THE INDUSTRIAL AGE, RIGHT UP TO THE PRESENT DAY!

OOOOOHHH!

BUT ADAMSKI'S GONE CRAZY! HE'S DIGGING HIS OWN CLIENTS IN DEEPER!

SILENCE IN COURT!

BANG! BANG! BANG!

MAY IT PLEASE THE COURT, I WOULD LIKE TO CALL MY CLIENTS BACK IN.

REQUEST GRANTED.

CALL THE ACCUSED!

BANG! BANG!

SO, YES, LADIES AND GENTLEMEN OF THE JURY, MY CLIENTS ARE MEMBERS OF THE SAPIENS GANG, BUT NEED I REMIND YOU THAT SAPIENS...

IS ALL OF US!!!

YOU WANTED TO FIND THE OTHER MEMBERS OF THE GANG... AND HERE THEY ARE!

EVERY ONE OF US IN THIS COURTROOM IS AS GUILTY AS MY CLIENTS.

MAY I REMIND YOU THAT MY CLIENTS HAVE BEEN HELD IN JAIL BY DETECTIVE LOPEZ FOR SOME TIME NOW, BUT THAT HASN'T STOPPED THE KILLING.

YOU NEED ONLY TAKE A LOOK AT WHAT'S HAPPENING IN OUR OCEANS RIGHT NOW.

UNLIKE THEIR LAND-LIVING COUSINS, LARGE SEA ANIMALS SUFFERED RELATIVELY LITTLE FROM THE COGNITIVE AND AGRICULTURAL REVOLUTIONS.

BUT MANY OF THEM ARE ON THE BRINK OF EXTINCTION NOW AS A RESULT OF THE INDUSTRIAL REVOLUTION AND HUMANS OVERUSING OCEAN RESOURCES! IF THINGS CONTINUE AT THIS PACE, IT'S LIKELY THAT WHALES, SHARKS, TUNA AND DOLPHINS WILL FOLLOW DIPROTODONS, GROUND SLOTHS AND MAMMOTHS INTO OBLIVION.

AMONG THE WORLD'S LARGEST CREATURES, THE ONLY SURVIVORS OF THIS HUMAN FLOOD WILL BE HUMANS THEMSELVES, AND THE FARMYARD ANIMALS THAT SERVE AS GALLEY SLAVES IN NOAH'S ARK!

WE HAVE THE DUBIOUS DISTINCTION OF BEING THE DEADLIEST SPECIES IN THE HISTORY OF BIOLOGY!!!

WE'RE ALL GUILTY! AND IT'S TIME WE REALIZED THAT...

WHEN MY CLIENTS COMMITTED THEIR CRIMES, THEY WEREN'T AWARE OF THE CONSEQUENCES OF THEIR ACTIONS. BUT TODAY WE ALL UNDERSTAND. AND WE DO LITTLE MORE ABOUT IT THAN MY CLIENTS DID! I HOPE THIS WILL BE TAKEN INTO DUE CONSIDERATION WHEN YOU PASS SENTENCE. THANK YOU.

SILENCE! SILENCE!

THE JURY WILL WITHDRAW TO DELIBERATE.

SOME TIME LATER...

HUH-HUM...

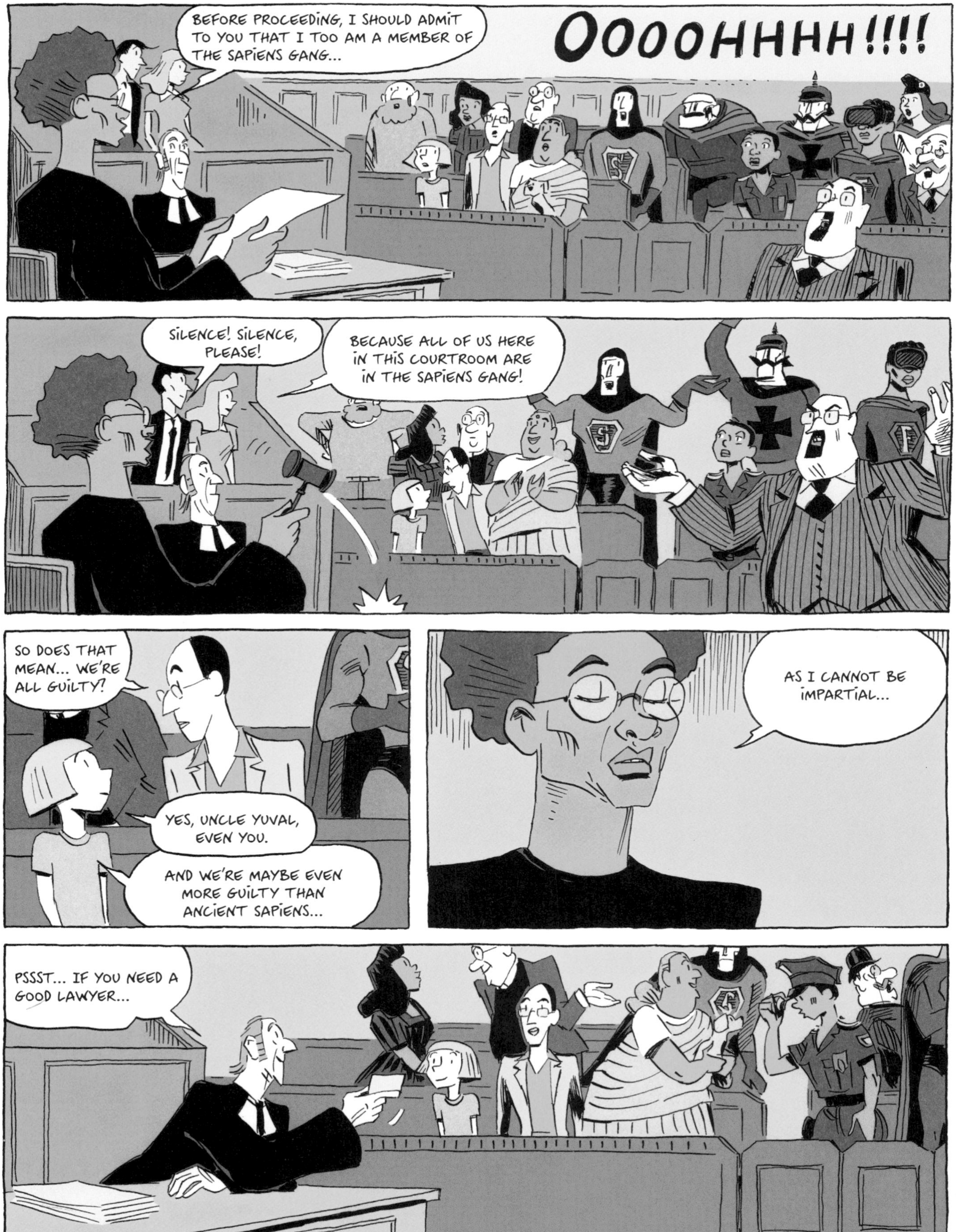
BEFORE PROCEEDING, I SHOULD ADMIT TO YOU THAT I TOO AM A MEMBER OF THE SAPIENS GANG...
OOOOHHHH!!!!
SILENCE! SILENCE, PLEASE!
BECAUSE ALL OF US HERE IN THIS COURTROOM ARE IN THE SAPIENS GANG!
SO DOES THAT MEAN... WE'RE ALL GUILTY?
YES, UNCLE YUVAL, EVEN YOU.
AND WE'RE MAYBE EVEN MORE GUILTY THAN ANCIENT SAPIENS...
AS I CANNOT BE IMPARTIAL...
PSSST... IF YOU NEED A GOOD LAWYER...

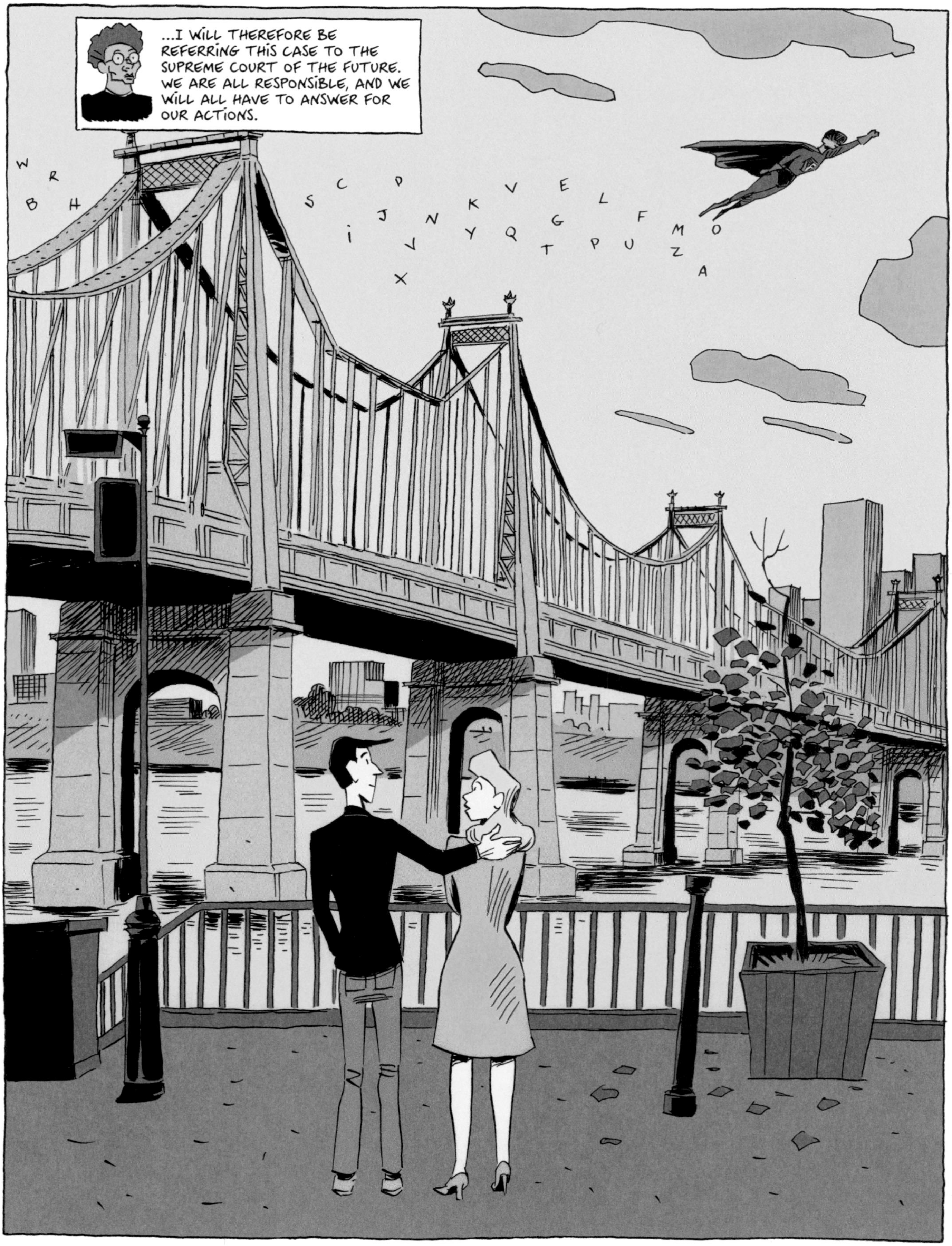
...I WILL THEREFORE BE REFERRING THIS CASE TO THE SUPREME COURT OF THE FUTURE. WE ARE ALL RESPONSIBLE, AND WE WILL ALL HAVE TO ANSWER FOR OUR ACTIONS.

ABOUT THE SPECIES OF THE GENUS HOMO

Many of the specific characteristics of the ancient Homo species are still being studied and debated. This does not mean that everything is debatable. It can be said with certainty that several species of humans existed in the past. They lived in Africa, Asia and Europe. Some of them became extinct, others interbred and evolved. Nowadays, there is only one Homo species, us: Homo sapiens. All contemporary humans throughout the world are Homo sapiens.

ALSO BY YUVAL NOAH HARARI

21 Lessons for the 21st Century

Homo Deus: A Brief History of Tomorrow

Sapiens: A Brief History of Humankind

ACKNOWLEDGMENTS

I would like to express my gratitude to the following people, without whom this project would not have been possible:

To David Vandermeulen and Daniel Casanave, who first came up with the idea of this graphic novel. Their creative genius has enabled the three of us to retell human history from a completely fresh perspective—I could not have done it without them. Their humor and intelligence made it a truly enjoyable experience. It has been sheer fun to work together.

To Martin Zeller for editing the French text and enabling our bilingual team to work together smoothly.

To Adriana Hunter, whose translation of the text from French to English not only bridged many cultural gaps but also helped bring the characters to life.

To Claire Champion for coloring and adding depth to the illustrations.

To Anne Michel, Lauren Triou, Aurelie Lapautre and the entire Albin Michel team who helped make this collaboration happen even in the midst of a global pandemic.

To Professor Robin Dunbar, whose ideas inspired my thinking for many years, and who kindly agreed to be featured as a character.

To Sapienship's CEO and CMO, Naama Avital and Naama Wartenburg—for their guidance, their dedication and their ideas, and for making a captivating story emerge out of a sea of endless details—and to Daniel Taylor for ensuring that story gets printed worldwide.

To the rest of the Sapienship team: Shay Abel, Michael Zur, Guangyu Chen, Hannah Morgan, Nina Zivy, Jason Parry, Galiete Gothelf, Erick Marchan and as a research assistant, Katia Zotovski, for their enthusiastic and professional work, whether in a cramped Tel Aviv apartment, in shiny new offices or locked down in their own homes.

To Slava Greenberg for his invaluable input on human diversity and for saving us from many pitfalls.

To my mother, Pnina Harari, for her continued support.

To Itzik Yahav, my husband and cofounder of Sapienship, for his vision and faith, and his ability to not only dream big but also realize big dreams.
—**YUVAL NOAH HARARI**

My thanks to Professor Robin Dunbar and Antoine Balzeau of the National Museum of Natural History in Paris.
—**DAVID VANDERMEULEN**

Thank you to my dear friends, Franck Bourgeron and Gilles Rochier.
Thank you to Christian Lerolle for his help.
—**DANIEL CASANAVE**

Thank you to Noé, Caroline and Franck for their valuable help.
—**CLAIRE CHAMPION**

David Vandermeulen, Daniel Casanave and the editor, Martin Zeller, would like to thank Yuval Noah Harari for agreeing to embark on this illustrated adventure with them, and the Sapienship team for their unfailing support and their enthusiasm.

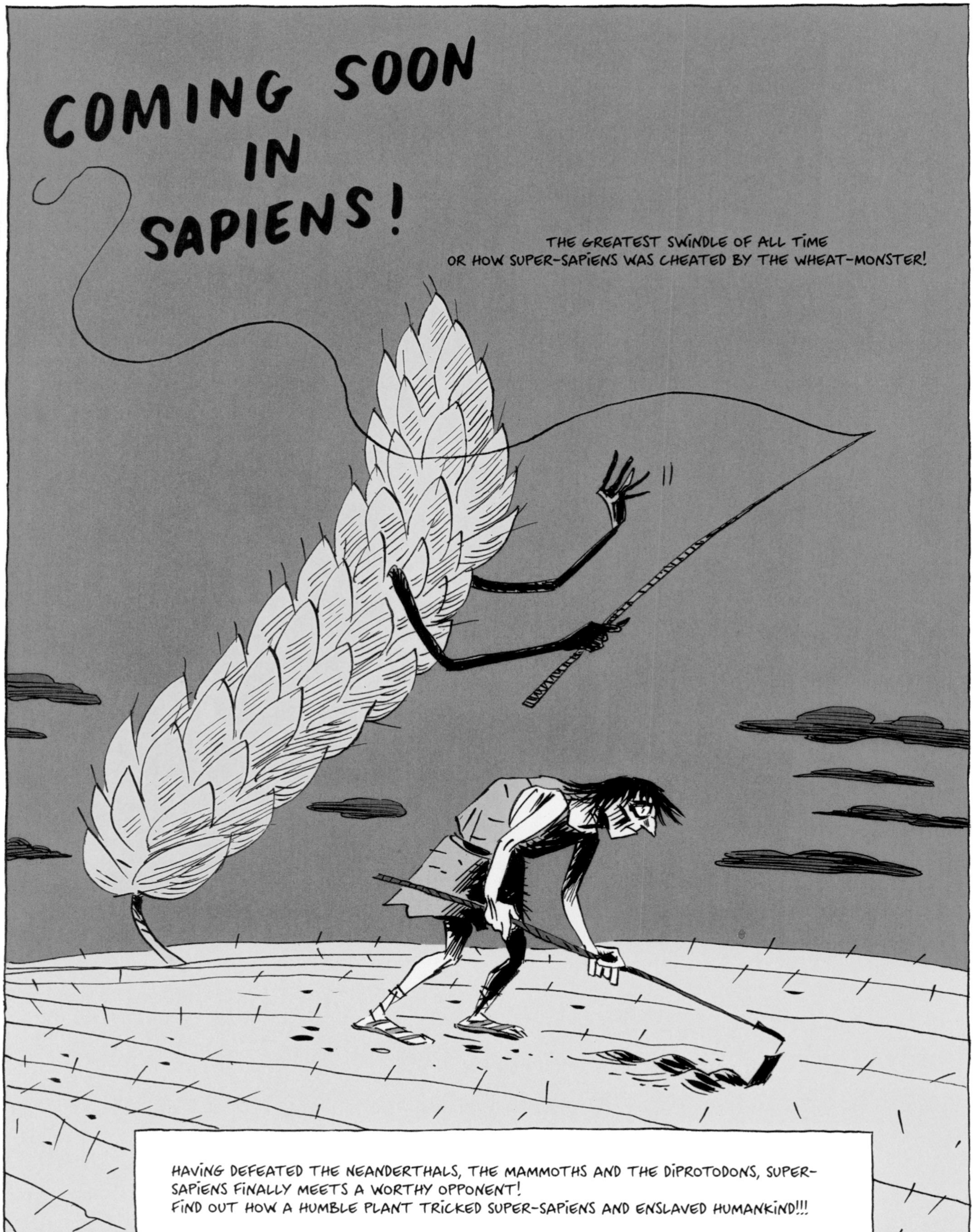
COMING SOON IN SAPIENS!
THE GREATEST SWINDLE OF ALL TIME
OR HOW SUPER-SAPIENS WAS CHEATED BY THE WHEAT-MONSTER!
HAVING DEFEATED THE NEANDERTHALS, THE MAMMOTHS AND THE DIPROTODONS, SUPER-SAPIENS FINALLY MEETS A WORTHY OPPONENT!
FIND OUT HOW A HUMBLE PLANT TRICKED SUPER-SAPIENS AND ENSLAVED HUMANKIND!!!